MAGNETIC BUBBLES

MAGNETIC BUBBLES

Edited by

H. JOUVE

*Laboratoire d'Electronique et de
Technologie de l'Informatique
Grenoble, France*

1986

Academic Press

Harcourt Brace Jovanovich, Publishers
London   Orlando   San Diego   New York   Austin
Boston   Sydney   Tokyo   Toronto

ACADEMIC PRESS INC. (LONDON) LTD
24/28 Oval Road, London NW1 7DX

United States Edition Published by
ACADEMIC PRESS INC.
Orlando, Florida 32887

Copyright © 1986 by
ACADEMIC PRESS INC. (LONDON) LTD

All rights reserved. No part of this book may be reproduced
in any form by photostat, microfilm, or any other means,
without written permission from the publishers

British Library Cataloguing in Publication Data

Magnetic bubbles.
1. Domain structure
I. Jouve, H.
538'.3   QC754.2.M336

Library of Congress Cataloging in Publication Data
Main entry under title:

Magnetic bubbles.

Includes index.
1. Magnetic bubbles. 2. Garnet. 3. Amorphous
substances. I. Jouve, H.
QC754.2.M34M34   1985   530.3'1   85-15015
ISBN 0-12-391220-2 (alk. paper)

Printed in Great Britain by Galliard (Printers) Ltd, Great Yarmouth

Contributors

C. H. Bajorek, IBM Thomas J. Watson Research Center, Yorktown Heights, P.O. Box 218, New York 10598, U.S.A.

C. D. Brandle, Bell Laboratories, Murray Hill, New Jersey 07974, U.S.A.

D. J. Breed, Philips Research Laboratory, 5600 MD Eindhoven, The Netherlands.

P. Chaudhari, IBM Thomas J. Watson Research Center, Yorktown Heights, P.O. Box 218, New York 10598, U.S.A.

U. Enz, Philips Research Laboratory, 5600 MD Eindhoven, The Netherlands.

M. H. Kryder, IBM Thomas J. Watson Research Center, Yorktown Heights, P.O. Box 218, New York 10598, U.S.A.

D. J. Muehlner, Bell Laboratories, 600 Mountain Avenue, Murray Hill, New Jersey 07974, U.S.A.

T. J. Nelson, Bell Laboratories, 600 Mountain Avenue, Murray Hill, New Jersey 07974, U.S.A.

S. Orihara, Fujitsu Laboratories Ltd, 1015 Kamikodanaka, Nakaharaku, Kawasaki, Japan.

T. Yanase, Fujitsu Laboratories Ltd, 1015 Kamikodanaka, Nakaharaku, Kawasaki, Japan.

Preface

Since its invention more than fifteen years ago, magnetic bubble technology has evolved considerably on the device level, aided by a large research effort in different laboratories all over the world.

The main characteristic of this technology has been the forging of a link between traditional magnetic mass memories using magnetic domains in order to store information and the fabrication processes developed for silicon integrated circuits. On the way, this emerging technology has had to find a market against competitors like semiconductor memories for small capacity applications and electromechanical magnetic memories for large capacity applications. As these two technologies have evolved considerably over the last ten years by bit density increase and cost reduction, so the magnetic bubbles must simultaneously face two tasks: establishing themselves as a reliable technology and following its competitors' tendency towards density increase and cost reduction. This dynamic situation has motivated a large research effort in order to increase magnetic bubble performances.

The purpose of this book is to describe some of the main subjects where outstanding results have been obtained, allowing further evolution of magnetic bubble technology.

One of the key points in garnet technology has been the ability to fabricate a monocrystalline substrate with an adequate lattice parameter for growing epitaxial magnetic garnet layers. The chapter by Dave Brandle on gadolinium gallium garnet wafers describes the research and results that have yielded high quality wafers suitable for epitaxy of magnetic films with a very low defect density and good surface quality suited for processing geometrical dimensions of the order of a micrometer.

Metallic films comprise another class of magnetic materials containing bubble domains: among these, some of the most exciting class are amorphous films of rare earths/transition metals alloys. The paper by Chaudhari *et al.* gives a synthetic description of the physical characteristics of this class of material, as well as their main properties as bubble materials.

One of the most important features of magnetic bubble domains is their low coercivity and high mobility. The chapter by Breed and Unz describes a very interesting phase of research oriented towards high mobility materials. The garnet material family is so flexible that, by adequate tailoring, bubble compositions allowing a 10 MHz clock frequency could be achieved (nearly 100 times the ordinary value).

Pattern propagations for bubbles defined with small elements (asymmetric chevrons for example) of permalloy have given rise to successive generations of magnetic bubble devices with capacities ranging from 64 Kbits in 1976 up to 1–4 Mbits in 1984. This progress was possible through the use of a large number of improvements and innovations, among which some of the most significant are described by S. Orihara and T. Yanase.

Ion implantation into magnetic garnets is already used in present devices in order to allow for hard bubble suppression. For the future, it has the main advantage of permitting the realization of bubble circuits with tolerant geometries. This capability opens the way to the fabrication of 4 and 16 Mbit devices. The chapter by T. J. Nelson and D. J. Muehlner describes the present status of this technique, showing its physical and device implications.

Besides the authors of the different chapters, this book could not have been written without the large amount of work that has been performed by researchers from many institutions who have contributed to the rapid advance of magnetic bubble science and technology. Among them, I wish to acknowledge particularly the helpful contribution of Ray Wolfe (Editor, *Applied Solid State Science*) for establishing the content of this book.

H. JOUVE

Contents

Contributors

Preface

1 Garnet Substrates for Magnetic Bubble Films 1

   C. D. BRANDLE

   1 Introduction 1
   2 Lattice parameter mismatch 3
   3 Bulk crystal defects 4
   4 Point defects 10
   5 Crystal stoichiometry 18
   6 Other garnet compositions 20
     References 28

2 Amorphous Gd-Co Alloys for Magnetic Bubble Applications 31

   P. CHAUDHARI, C. H. BAJOREK and M. H. KRYDER

   1 Introduction and general background 31
   2 Film fabrication 34
   3 Atomic arrangement in amorphous Gd-Co alloys 41
   4 Magnetic moment 47
   5 Annealing and ion implantation 55
   6 Magnetic anisotropy 59
   7 Domain wall dynamics of amorphous films 66
   8 Amorphous film bubble devices 75
   9 Summary and conclusions 85
     References 86

3 Garnet Films for High Bubble Velocities and High Bubble   91
   Mobilities . . . .

   D. J. BREED and U. ENZ

   1 Introduction 92
   2 Garnet films with orthorhombic anisotropy 97
   3 Bismuth-containing garnet films with high uniaxial anisotropy and
     low damping 126

	4 Device implications.	130
	5 Conclusion	133
	References	134

4 Field Access Permalloy Devices 137

S. ORIHARA and T. YANASE

1	Introduction.	138
2	Problems in higher density devices	139
3	Design of function elements	143
4	Optimization of propagation	165
5	Loop organization.	178
6	Design and characteristics of a 1-Mbit chip.	184
7	Further development of permalloy devices	193
	References	210

5 Circuit Design and Properties of Patterned Ion-Implanted Layers for Field Access Bubble Devices 215

J. J. NELSON and D. J. MUEHLNER

1	Introduction.	216
2	Chip organization	224
3	Active components—transfer	234
4	Active components—generation, replication, and detection.	245
5	Propagation.	256
6	Charged walls	263
7	Damage and strain profiles in the implanted layer	271
8	Magnetic properties of the implanted layer.	277
	References	288

Subject Index 291

1 Garnet Substrates for Magnetic Bubble Films

C. D. BRANDLE

*AT & T Bell Laboratories
Murray Hill, New Jersey*

1 Introduction 1
2 Lattice parameter mismatch 3
3 Bulk crystal defects 4
   3.1 Facet formation and strain 5
   3.2 Growth striations 6
   3.3 Reduction of faceting and striations 8
4 Point defects 10
   4.1 Inclusions 10
   4.2 Dislocations 12
5 Crystal stoichiometry 18
6 Other garnet compositions 20
   6.1 Ion substitution 21
   6.2 Gallium garnets 22
   6.3 Solid solutions 23
   6.4 Scandium substituted garnets 24
   6.5 Coupled substitution 26
   References 28

1 INTRODUCTION

The use of rare-earth iron garnets as the active film in a bubble device requires the deposition of the film material on a suitable substrate. Two techniques have been developed to accomplish this goal: (1) chemical vapor deposition (CVD) (Mee *et al.*, 1969; Heinz *et al.*, 1972a) and (2) liquid phase epitaxy (LPE) (Levinstein *et al.*, 1971; Blank *et al.*, 1973). For either process, the substrate requirements remain the same:

(1) It must be nonmagnetic, i.e. it cannot interact with the magnetic properties of the film.

(2) The lattice parameter of the substrate must be in the range accessible to that of the film and, just as important, must be uniform from wafer to wafer and through the length of the crystal.

(3) The substrate must have a high degree of internal perfection, i.e. low dislocation density, low strain and inclusion free.

(4) The substrate must be able to be fabricated with a defect free surface finish, uniform edge and be flat to within $3\,\mu$m.

Work on the growth and perfection of the substrate material has been centered around gadolinium gallium garnet (GGG), but has been extended to cover other garnet compositions as well, for the problems associated with the growth of GGG are common to all other possible garnet substrates.

Because of the scope of this review, only the first three items mentioned will be covered in detail since surface perfection is strongly dependent upon material processing rather than material perfection. This does not imply, however, that poor substrates can yield a good quality surface, but that the rigid bulk material requirements preclude the type of surface imperfections usually associated with inclusions and dislocations.

The development and perfection of the LPE technique for garnet film deposition produced the advantage of lower growth temperature and much more flexibility of cation substitution in the film, thereby allowing an exact lattice parameter match of the film to the substrate. At this point, it quickly became recognized that the internal quality of the substrate would play a dominating role in the perfection and use of the film. Thus, intensive efforts were begun by numerous investigators to determine the origin of substrate/film defects and seek their elimination.

Unlike the rare-earth iron garnets and some of the rare-earth aluminum garnets, the rare-earth gallium garnets have a congruent melting composition, i.e. they do not decompose upon melting. It is therefore possible to utilize the Czochralski technique to grow large single crystals of these compounds for substrate use. Linares (1964) first reported on the Czochralski growth of gadolinium gallium garnet. In his paper he noted several difficulties, the primary problem being the vaporization and subsequent loss from the melt of the more volatile gallium oxide. Since that time, this problem and those associated with the growth of low dislocation, low strain GGG have been largely solved and today GGG crystals are being grown commercially which yield fabricated substrates 4.0 in. (100 mm) in diameter with dislocation densities of less than $0.5/\text{cm}^2$. A summary of the conditions used to grow

Table 1 Growth conditions for substrate garnet materials.

Condition	$Gd_3Ga_5O_{12}$	$Ln_3Sc_2Ga_3O_{12}$	$Ln_3Sc_2Al_3O_{12}$
Melt composition	Nonstoichiometric	Nonstoichiometric	Nonstoichiometric
Atmosphere	N_2–2% O_2	N_2–2% O_2	N_2
Pull rate	5–8 mm/hr	6–7 mm/hr	4–5 mm/hr
Rotation rate	20–50 rpm	30–50 rpm	5–15 rpm (>85 for flat interface)
Diameter	3.8–10.0 cm	1.5–3.8 cm	1.5–3.8 cm
Length	10–35 cm	5–10 cm	5–10 cm

GGG as well as other garnets which will be discussed later is given in Table 1. The conditions shown will vary somewhat depending upon such factors as the furnace geometry, crucible size, melt composition and position of the crucible in the RF coil; however, they are representative of those used to produce high-quality substrate material.

2 LATTICE PARAMETER MISMATCH

One of the first considerations for selection of a substrate material which would produce good epitaxial growth of rare-earth iron garnet films is the lattice parameter of the film and its "fit" on the substrate being used. When one considers the lattice parameter range of the iron garnets, 12.283 Å for $Lu_3Fe_5O_{12}$ to 12.529 Å for $Sm_3Fe_5O_{12}$ (Geller, 1967), all rare-earth aluminum garnets can be immediately ruled out as substrates because of their small lattice parameter. This then leaves three other classes of nonmagnetic garnets from which to choose the substrate:

(1) Rare-earth gallium garnets ($Ln_3Ga_5O_{12}$) and their solid solutions.
(2) Scandium substituted rare-earth gallium and aluminum garnets ($Ln_3Sc_2Ga_3O_{12}$ and $Ln_3Sc_2Al_3O_{12}$).
(3) Rare-earth gallium garnets substituted with ions other than trivalent, e.g. Ca^{2+}, Mg^{2+}, Zr^{4+}.

Because of the close ionic size of Fe^{3+} and Ga^{3+}, 0.645 Å and 0.62 Å respectively for octahedral coordination and 0.49 Å and 0.47 Å respectively for tetrahedral coordination (Shannon and Prewitt, 1969), there exists considerable overlap of the lattice parameters for the rare-earth iron and rare-earth gallium garnets. This lattice parameter overlap does not exist to as great an extent for the two other classes mentioned. Therefore, of the three classes, the rare-earth gallium garnets have the most desirable lattice parameters and properties for Czochralski growth. It is in this class where most of the work has been done.

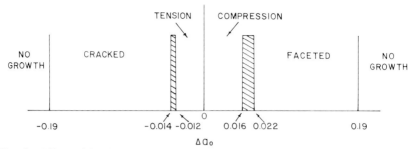

Fig. 1 Effect of lattice parameter mismatch on film morphology. (From Blank and Nielsen, 1972.)

In the CVD process for film growth, lattice mismatch between the film and the substrate is extremely important. At the deposition temperature (approximately 1100–1200°C), the only source of the magnetic anisotropy is due to the stress induced in the film because of lattice mismatch. In the LPE process, the source of magnetic anisotropy is not predominantly stress induced, but is growth induced, i.e. it depends upon a partial ordering during growth of the rare-earth ions on the dodecahedral crystallographic site (Rosencwaig et al., 1971). Therefore, large stresses in LPE films are usually avoided by adjusting the film composition for an exact lattice parameter match, whereas in CVD films, stresses are necessary to produce the desired anisotropy. Blank and Nielsen (1972) have determined the limits of lattice mismatch. Films deposited with a high compressive stress yielded rough faceted surfaces whereas those films deposited with a high tensile stress usually showed cracks. Their results are shown graphically in Fig. 1.

Miller and Caruso (1974) have examined highly compressive films and determined that the "faceting" observed by Blank and Nielsen (1972) on the surface of the film is the result of the presence of dislocations (10^7–10^8/cm^2) introduced due to lattice mismatch. Thus, one of the first requirements for the substrate is that its lattice parameter be in the range which is accessible to the rare-earth iron garnet film and does not exceed the limits shown in Fig. 1. As already mentioned, the rare-earth gallium garnets are in this range, and gadolinium gallium garnet falls midway between the two lattice parameter extremes for the rare-earth iron garnets.

3 BULK CRYSTAL DEFECTS

Bulk crystal defects can be classified as those which can occur either completely across a section of the crystal or throughout its entire length. Usually these defects are a result of small lattice parameter changes within the bulk crystal. These lattice parameter changes result in internal strains

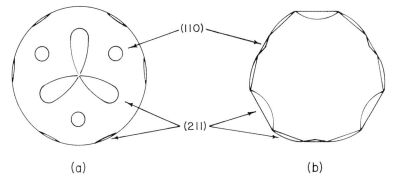

Fig. 2 Schematic representation of facets formed on the growth interface of (111) grown crystals with (a) slow rotation or (b) fast rotation.

within the crystal which can be replicated in the epitaxial film after the crystal is fabricated into substrates (Glass et al., 1973). It is therefore necessary to reduce to a minimum the growth induced strain in the substrate so that high quality epitaxial films can be produced.

3.1 Facet formation and strain

Facet formation and the resulting strain usually occurs on the solid–liquid interface when an area of the interface becomes tangent to a direction of naturally occurring facets. For garnets, the naturally occurring facets are the [110] and [211] planes and therefore a [111] crystal will have two distinct areas of facet formation (Basterfield et al., 1968). There will be a central area or core composed of three (211) facets and a middle area composed of three additional (110) facets. A third area, again composed of (211) and (110) facets, can also be present near the surface of the crystal. These areas are shown schematically in Fig. 2a.

At the present time, the origin of the facet strain is not completely understood; however, three sources have been suggested (Cockayne et al., 1973): (1) impurity segregation, (2) changes in stoichiometry, i.e. in the Gd/Ga ratio, and (3) oxygen vacancies. Typically, starting materials for GGG growth consist of material with a minimum 4–9s purity specification. Typical non-rare-earth impurity levels are less than 50 ppm while the remainder are rare-earth impurities. If one assumes a distribution coefficient of approximately 1 and that Vegard's Law applies, then the 50 ppm rare-earth impurity level would result in $\Delta a \approx 5 \times 10^{-4}$; however, the magnitude and sign of this difference would be dependent upon the specific rare-earth impurity, whereas it has been reported by several workers (Cockayne et al.,

1973; Glass, 1972) that the lattice parameter is larger in the faceted region than in the matrix.

Glass (1972) has proposed a model based on a variation of the Gd/Ga ratio between the faceted region and the matrix. His argument for a Ga rich facet region was based on the assumption that stoichiometric GGG has a lattice parameter of 12.383 Å. This value has been shown to be the lattice parameter of a Gd rich composition $Gd_{3.03}Ga_{4.97}O_{12}$ (Geller et al., 1972; Carruthers et al., 1973; Brandle and Barns, 1974). Using his reported value (Glass, 1972) of $\Delta a = 10^{-3}$ Å, the change in Gd/Ga ratio necessary to produce this lattice parameter variation would be about 0.1%.

Cockayne et al. (1973) have compared the lattice parameters of the faceted regions, (211) and (110), with those of the matrix and have found that the facet region lattice parameter exceeds that of the matrix ($\Delta a \approx 1.3 \times 10^{-3}$ Å), in agreement with Glass. They also reported this same difference was observed in YAG and suggested that the same mechanism of strain was present in both crystals. Because of this observation, they concluded that the strain in the faceted region was due to an oxygen segregation effect, i.e. oxygen vacancies in the faceted region. Stacy et al. (1974a) and Stacy (1974), using fluorescence and thermoluminescence techniques, have reported that the faceted region of GGG crystals contain a higher concentration of electron traps, i.e. oxygen vacancies, than the matrix. These results are in agreement with the above conclusions.

The oxygen vacancy model for facet strain is also supported by the facet lattice parameter measurements which in all cases have been reported to be larger than the matrix. Metselaar and Huyberts (1973) have shown that in the case of polycrystalline YIG, the formation of oxygen vacancies compensated by ferrous ions increases the lattice parameter by as much as 0.003 Å. The amount of increase depends upon the oxygen partial pressure at the firing temperature. Furthermore, they found that "pure" YIG contains an intrinsic ferrous content. In the case of GGG, low-level impurities or cation vacancies could provide the compensation necessary for oxygen vacancy formation.

3.2 Growth striations

Growth striations are common in many crystals (Laudise, 1970) and can be seen by numerous techniques (Witt et al., 1973; Schwuttker, 1962; Kikuta et al., 1966). They are the result of rapid fluctuations in the instantaneous growth rate of the crystal due to random or periodic temperature changes at the solid–liquid interface or changes in the pulling rate (Morizane et al., 1967). A typical set of striations in a GGG crystal is shown in Fig. 3 (Miller,

1973). The crystal was cut lengthwise parallel to the (111) growth direction. The facet region is to the right of the photograph, while the nonfacet region is at the left. The fine periodic striations are rotational in origin and due to a nonsymmetric temperature distribution about the rotational axis of the crystal, whereas the widely spaced striations are due to large random temperature excursions causing considerable remelt of the growth interface. Note that the striations also appear in the faceted region of the crystal. Belt

Fig. 3 Growth striations is a (111) GGG crystal longitudinally cut and etched in hot orthophosphoric acid. The non-faceted region is to the left and faceted region to the right. (From Miller, 1973, with permission of the Electrochemical Society, Inc.)

and Moss (1973), using electron microprobe scans, have identified the origin of these striations as due to a variation in the Gd/Ga ratio from that of the surrounding area.

Such compositional variations are also reflected in lattice parameter changes and low-level strain (Belt and Moss, 1973) across the striation. These variations can be replicated in the epitaxial layer (Glass *et al.*, 1973; Stacy *et al.*, 1974a; Keig, 1973) and impede domain motion through interaction with the magnetic film properties. Basterfield *et al.* (1968) have shown that the strain associated with these striations is normal to the striation, and therefore the degree of strain transmitted to the film depends upon the angle of intersection of the striations with the substrate surface.

3.3 Reduction of faceting and striations

As indicated in the previous section, the strain associated with faceting on the growth interface or striations can be detrimental to the performance of a magnetic bubble device. Therefore, the elimination or at least a reduction in magnitude of these defects is necessary to produce usable substrate material.

Facet strain can be reduced or eliminated from the substrate by either (1) completely suppressing facet formation by selection of an appropriate growth direction or (2) changing the position of the facets on the growth interface so the resulting strain can be removed during fabrication. This can be accomplished by changing liquid thermal gradients, thereby altering the shape of the solid–liquid interface. The tangent point for facet formation moves out across the crystal radius as the interface becomes less convex to the melt. For the garnet system the first choice is not possible because of the high symmetry associated with the [211] and [110] facets; therefore the second choice must be considered.

Cockayne *et al.* (1968) have shown that under the proper conditions, the shape of the growth interface for YAG can be controlled by the rotation rate of the crystal. At sufficiently high rotation rates (150 rpm) he was able to grow YAG with a very shallow or flat interface which was completely free of facets and strain in the central area of the crystal. Using his technique, Brandle and Valentino (1972) were able to grow GGG with a "flat" interface at rotation rates of 25 to 50 rpm. The lower rotation rate was required because of the lower thermal conductivity of GGG as compared to YAG. It should be pointed out that the facets on the growth interface were not eliminated by this procedure, but simply moved from one area of the crystal to another. The strain associated with the facets is now on or near the surface of the crystal and can easily be removed during the fabrication step, yielding a strain-free central section which can then be processed into substrates (Fig. 2b). From the above discussion, it is apparent that the control of facet formation for Czochralski grown garnets can only be accomplished during the growth process and is strongly dependent upon the furnace geometry and growth conditions.

The transition from a conical growth interface to a flat interface is the critical process in the growth of garnet substrate materials, for this transition represents a significant change in the thermal geometry and flow structure of the molten garnet oxide. Melt simulation experiments (Brandle, 1977; Miller and Pernell, 1981, 1982) to model GGG melts have shown that two distinct regions of flow are formed in the liquid when the interface transition occurs. The fluid in the region around the crystal is driven by the rotating crystal interface which is acting as a "centrifugal pump". This pumping

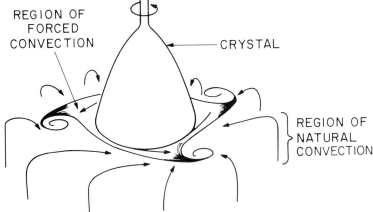

Fig. 4 Schematic representation of a GGG melt surface showing areas of forced and free convection and eddy swirls.

action draws hot liquid from the center of the crucible and forces it across the surface. The second region near the crucible wall is dominated by natural convection and tends to flow radially from the crucible wall towards the center as shown schematically in Fig. 4.

Carruthers (1976) has shown that this transition is dependent upon the rotation rate of the crystal, the crystal diameter and the temperature difference in the liquid. Miller *et al.* (1978) and Brandle (1978) have shown that most dislocations present in bulk GGG crystals occur during the interface transition and are a direct result of growth instabilities during the remelt process. Takagi *et al.* (1976) have found that the rotation rate required to produce a flat interface is also dependent upon the pulling rate—the faster the growth rate, the slower the rotation rate. They have also confirmed the change in the fluid flow by measuring the temperature change at the bottom of the crucible during the time when the interface goes from convex to flat. Recently, Brandle (1982) has shown that this transition is a form of baroclinic instability and is common in many oxide melts.

As already mentioned, the second type of bulk defect which can be detrimental to device performance is striations. Striations can be considerably reduced in magnitude by suppressing thermal oscillations in the liquid, either by reducing the temperature gradients (Zupp *et al.*, 1969) or by using baffles (Whiffin and Brice, 1971) in the melt. Reducing the temperature gradient makes crystal growth more difficult and promotes facet formation while baffles reduce the usable volume of melt. An alternate approach to the use of baffles is to use a crucible with a height to diameter ratio (H/D) of less than one—ideally using a value of $\frac{1}{2}$ to $\frac{1}{3}$ (Whiffin and Brice, 1971). This produces crystals with fewer striations but as in the low gradient technique, growth is more difficult. A similar approach is to use a normal crucible $H/D = 1$, but to have a liquid height of only $\frac{1}{2} H$. Crystals grown under these conditions also have fewer striations but they are much more difficult to start (C. D. Brandle and D. C. Miller, 1973 unpublished observations). This alternate method is approached under normal growth conditions when approximately 60–70% of the liquid is "pulled" from the crucible. The lower sections of these crystals do show less pronounced striations than do the upper sections (Miller et al., 1978). Because the strain associated with growth striations is normal to the striations, another approach would be to grow the crystal so that the striations would be parallel to the substrate surface. The angle of intersection between the striations and the substrate surface would then be very small, and little, if any, strain would be transmitted to the film. This can be accomplished by growing the crystal with a flat interface as discussed above.

4 POINT DEFECTS

It has been well established that point defects in magnetic garnet films can interact with and impede domain motion. These point defects can be divided into two broad categories: (1) those caused by inclusions within the substrate and (2) those related to lattice defects such as dislocations and interstitial ions. Inclusions can have several sources of formation, e.g. contamination from the crucible material, an insoluble second phase in the liquid, or precipitation of another phase from solid solution as the crystal cools. Each of the above also can cause dislocations and strain. In addition to the above sources, dislocations can also be generated from improper melt stoichiometry and poor thermal environment which leads to rapid, uncontrolled growth at the solid–liquid interface.

4.1 Inclusions

The effect on the film of inclusions in the substrate can be through interaction with dislocations generated by the inclusion or for an inclusion that intersects the surface, the formation of a hole or region of highly dislocated material in the film directly above the inclusion. In either case, such defects limit the usable area of the film. For the case of gadolinium gallium garnet crystals, inclusions have been identified as (1) small hexagonal or triangular platelets of iridium (O'Kane et al., 1973; Brandle et al., 1972), (2) small, opaque cubic particles (Brandle et al., 1972), and (3) needle-shaped transparent inclusions (Brandle et al., 1972).

Iridium inclusions are a direct result of contamination of the liquid with the crucible material. Several mechanisms have been proposed for the transport of iridium from the crucible to the liquid and can be summarized as follows. The first source is due to crucible cleaning. Usually one of the final steps in cleaning is to grit blast the crucible followed by a wash in an acid bath and then water. Small iridium particles which had been formed during the cleaning could then contaminate the liquid. Brandle and Valentino (1972) found that firing the crucible at 1800°C for 15 min prior to charging eliminated this source of contamination.

A second source is believed to be due to improper drying of the starting materials and/or moisture in the growth atmosphere. Water reacts with hot iridium producing the volatile IrO_2 which can then deposit in the cooler regions of the charge. Later, as the temperature is raised to the melting point, the IrO_2 decomposes back to metallic iridium.

A third source, first identified by Linares (1964), is due to the decomposition of Ga_2O_3 to form the suboxide and oxygen. Again the volatile iridium oxide is formed, providing a transport medium for the formation of metallic iridium. Elimination of this source of contamination can be accomplished by changing the atmosphere composition (Brandle et al., 1972) or prereacting the powders (M. Ishii, 1974 personal communication).

Small, opaque cubic inclusions were first reported by Brandle et al. (1972) and their effect on the substrate quality by Miller (1973). It has been proposed (O'Kane et al., 1973) that the cubic inclusions are a rare-earth gallium suboxide which is formed in the melt ahead of the growth interface. Evidence to support this mechanism was based on several facts: (1) strain or dislocations were rarely observed associated with the cubic inclusions indicating entrapment rather than solid–solid precipitation, (2) the elimination of these inclusions could be accomplished by growth in a slightly oxidizing atmosphere and (3) the rate of decrease in the precipitate density was proportional to the oxygen partial pressure in the growth atmosphere. Furthermore, they report a reduction in Ga_2O_2 loss when growth was carried out in an oxidizing atmosphere.

O'Kane et al. (1973) have reported that growth in a neutral atmosphere (N_2) did not cause suboxide inclusions from the decomposition of Ga_2O_3 and that no significant loss of Ga_2O_3 was observed from the melt. Similar results have been reported by M. Ishii (1974 personal communication). In this case, the constituent oxides were prereacted prior to being placed in the crucibles, whereas in the work reported by Brandle et al. (1972), unreacted oxides were used. This would suggest that the formation of this type of precipitate is strongly dependent upon initial material preparation.

The third type of inclusion reported was the clear needle form (Brandle et al., 1972). These are by far the least common and comprise only about

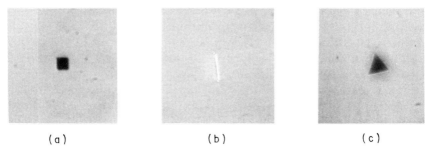

Fig. 5 Types of inclusions observed in $Gd_3Ga_5O_{12}$ substrates: (a) iridium, (b) clear needle type, (c) cubic opaque type. (From Brandle *et al.*, 1972, with permission of the North Holland Publishing Co.)

3-5% of the total inclusions in any given crystal. It is believed that these inclusions are Gd_2O_3 which has precipitated from the liquid due to a local supersaturation of that component at the solid–liquid interface. They were always found associated with the dark cubic type of inclusions. The elimination of this form of inclusion can be accomplished by growth in an oxidizing atmosphere (Brandle *et al.*, 1972). This suggests a close chemical relationship between the formation of this inclusion and the opaque, cubic form. Examples of each of these types of inclusions are shown in Fig. 5.

4.2 Dislocations

As mentioned above, dislocations in the magnetic garnet films can impede domain motion (Cockayne and Roslington, 1973). These dislocations have two sources: (1) those generated during the film growth and (2) those which are propagated into the film from the substrate. It is this second source which will be covered in detail. Dislocations in gallium garnet films can be divided into three major groups: (1) linear, (2) helical, and (3) closed circular loops.

4.2.1 Linear dislocations

Linear dislocations in gallium garnets have been observed by numerous authors using optical, chemical, and x-ray techniques (Miller, 1973; O'Kane *et al.*, 1973; Cockayne and Roslington, 1973; Stacy *et al.*, 1974b; Kishino *et al.*, 1974). When examined under crossed polarized light, they exhibit microscopic strain birefringence and are usually propagated almost parallel to the growth axis of the crystal. They can usually be traced back to the crystal-seed interface, a growth instability (Miller *et al.*, 1978), or the point in the crystal where the interface went from convex to flat (Miller et al., 1978, Brandle, 1978; Cockayne and Roslington, 1973). In any case, the distribution of dislocations in the cross section of the crystal can give a clue as to the specific cause.

Dislocations generated at the seed-crystal interface can be eliminated by using the technique reported by Dash (1959) and used by O'Kane et al. (1973) which involves a diameter reduction prior to a diameter increase. An alternate approach is to use dislocation free seeds and start growth of the crystal at the seed diameter. Another approach reported by Cockayne and Roslington (1973) is a facet blocking technique. They found that facets block the formation of dislocations and that the propagation of dislocations is strongly dependent upon the shape of the crystal interface. Thus, the shape of the growth interface determines the interface area covered by facets, thereby blocking dislocation formation in that area. By controlling the interface shape through rotation rate adjustments, they were able to prevent dislocation formation during the critical "shouldering" process. The convex interface was then remelted by a sudden increase in the rotation rate to produce the desired flat growth interface. Also, this process allowed the strain associated with the initial growth to be propagated outward toward the surface of the growing crystal.

Tominaga et al. (1974) have confirmed the above results by growing $Sm_3Ga_5O_{12}$ on GGG seeds. They found that the dislocations tend to propagate normal to the growth interface. For a convex interface, dislocations would tend to grow to the crystal surface while they would tend towards the center of the crystal for a concave interface. Therefore, the ideal interface shape is very slightly convex so to produce facets on the crystal surface and propagate dislocations away from the center. If a concave section of the interface is produced during the critical remelting of the conical growth interface, dislocations ($>500/cm^2$) can be generated because of impurity entrapment and segregation (Brandle, 1978; Cockayne and Roslington, 1973). Evidence of this interface shape can be seen as a heavy strained, dislocated section near the center of the crystal (Fig. 6a).

An alternate pattern of dislocation which can occur is characterized by a clear, low dislocation central region surrounded by a region of very high dislocation density near the surface of the crystal. In this case, rapid growth and impurity entrapment has occurred during the shouldering process and the dislocations are propagated parallel to the growth axis once a flat interface has been established (Fig. 6b).

A third pattern of dislocation has been reported (Brandle, 1978) and consists of a complete ring around the circumference of the crystal. This pattern was typically found in crystals having a diameter much larger than 0.5 times the crucible diameter. In this case, the strain associated with the

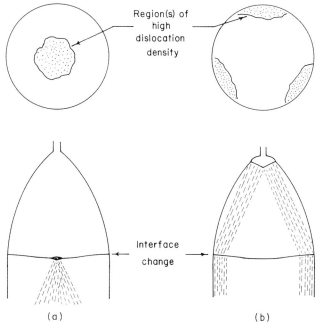

Fig. 6 Schematic representation of dislocation formation during (a) interface remelting and (b) seeding and "shouldering" process.

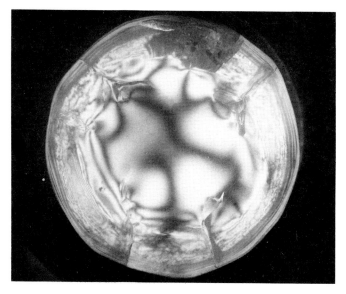

Fig. 7 Ring dislocation structure caused by fluid flow transition.

dislocations usually causes the surface of the crystal to fracture. A typical crystal having this dislocation pattern is shown in Fig. 7. The formation of these dislocations is a direct result of a second transition in the fluid after the interface remelt transition as discussed earlier. This second transition represents the complete collapse of the natural convection cell and the total domination of the fluid flow by forced convection. As indicated earlier, the exact nature of the interface remelt process is strongly dependent upon the ratio of crystal diameter/crucible diameter, the length of the crucible, the "pull" rate and the thermal geometry produced by a given furnace configuration.

4.2.2 Helical defects

A second form of defects which has been observed in Czochralski-grown GGG are large helical line defects (Stacy and Enz, 1972). These defects, unlike those described above, are much larger, typically from 150 to 300 μm in diameter, and have been observed in crystals covering a wide range of crystal perfection. The orientation is usually parallel to the growth axis, while the pitch of the helix shows a definite correlation with the growth striae of the crystal. The interior of the helix consists of a fine structure of lines radiating outward from the center which may or may not contain fine precipitates. Thus, the structure of the helix resembles that of a spiral staircase with a pitch corresponding approximately to the spacing between growth striations (Stacy and Enz, 1972). X-ray topographs have shown that (1) the dislocation helix is of the prismatic type, i.e. the Burgers vector is parallel to helix axis and (2) a compressive stress exists along the axis of the helix which implies an interstitial type of dislocation (Stacy and Enz, 1972; Matthews et al., 1973).

Matthews et al. (1973) have attributed the formation of these helices to a nonequilibrium point defect (Ga interstitials) based on the observation that excess Gd_2O_3 in the melt tends to suppress their formation. Nes (1974) however, has proposed a mechanism by which nucleation and growth of precipitates occur on the dislocation by diffusion of vacancies resulting in an interstitial climb loop. Data from Brandle et al. (1972) also indicate a decrease in the number of inclusions in crystals grown from Gd_2O_3 rich melts. As Stacy (1974) has pointed out, the model based on precipitation of a second phase also accounts for the radial "steps" seen in the helical defects and the observed lattice compression.

An example of this type of defect is shown in Fig. 8. Figure 8a shows this defect under transmitted light. In this case, light scattered from the precipitates can be seen to align along the "steps" of the spiral. Figure 8b shows the same defect after etching where it intersects the surface of the

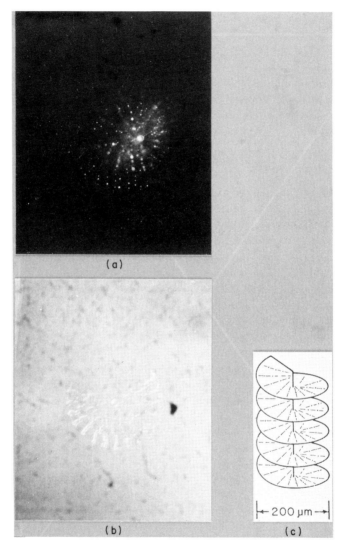

Fig. 8 Helical dislocation in GGG. (a) Transmitted light showing precipitate inclusions along radial "steps". (b) Surface of an etched wafer again showing radial "steps". (c) Schematic three-dimensional representation of a helical dislocation.

Garnet Substrates for Magnetic Bubble Films 17

polished wafer. Note that only one-half of the defect is etched because of its spiral nature. Figure 8c shows schematically the three-dimensional structure of the dislocation in the bulk material.

4.2.3 Circular loops

A third form of dislocations which has been reported (Morizane *et al.*, 1967; Stacy and Enz, 1972) is of the same type as the helical, but forming closed loops rather than an open helix. Thus, this form of defect can be considered as an isolated point defect and can vary from substrate to substrate. However, unlike the line dislocation discussed above, the closed loops can be from 5 to 200 μm in diameter (Matthews *et al.*, 1973; Lal and Mader, 1976) and usually consist of three loops of dislocations. Matthews *et al.* (1973) have reported that the dislocation loops are associated with a central inclusion

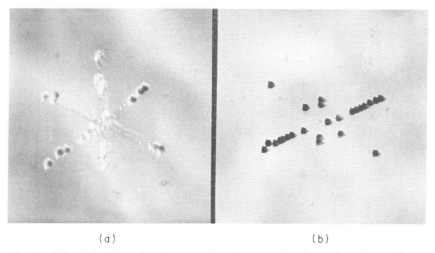

(a)   (b)

Fig. 9 Spherical dislocation loops (prismatic type) showing orientation of loops along (110) planes. (From Miller, 1973, with permission of the Electrochemical Society, Inc.)

and the {110} planes pass either through or extremely close to the inclusion. An example of this type of defect can be seen in Fig. 9 for a crystal grown along the (111) axis.

As in the case of the helical dislocations, Miller (1973) has reported that the loops can be observed with and without a precipitate. When the precipitate is present, the inclusions lie in the plane of the loop and not along the line of the loop. This observation would suggest a similar mechanism of formation as in the case of the helical defects.

5 CRYSTAL STOICHIOMETRY

Because of the importance of lattice parameter match between the substrate and the film, one of the first questions concerning the use of GGG as a substrate was the reproducibility of the lattice parameter of Czochralski-grown crystals.

In early work, flux-grown GGG crystals had been used as substrates for YIG (Linares, 1968) films. Gadolinium gallium garnet was chosen because its lattice parameter, 12.376 Å, has an exact match of that of YIG (Geller *et al.*, 1969). However, lattice parameter measurements on early Czochralski material showed a lattice parameter of about 12.383 Å for crystals grown from stoichiometric melts (Brandle *et al.*, 1972; Belt *et al.*, 1973). Such a variation in lattice parameter would preclude the use of that material as a substrate. To determine the range of variability in the lattice parameter of Czochralski-grown crystals, Brandle *et al.* (1972) grew a series of crystals ranging from 35.7 to 39.4 mole % Gd_2O_3. These crystals showed a lattice parameter variation from 12.381 Å for Ga-rich melts to 12.385 Å for Gd-rich melts. Ceramic samples prepared by Carruthers *et al.* (1973) showed lattice parameter variations from 12.376 Å for stoichiometric and Ga_2O_3 rich material to 12.411 Å for Gd_2O_3-rich material, indicating considerable solid solution of Gd_2O_3 in $Gd_3Ga_5O_{12}$.

Geller (1967) has reported similar results for the Y_2O_3–Ga_2O_3 system, while Suchow *et al.* (1970) and Suchow and Kokta (1972) have reported substitution of rare-earth ions on the octahedral site in gallium garnets. Thus, it appears that all Czochralski-grown GGG crystals are Gd-rich and can be represented by the general formula

$$Gd_3[Gd_xGa_{2-x}]Ga_3O_{12}. \tag{1}$$

The degree of substitution x on the octahedral site can be related to a change in the lattice parameter by using either the method of Carruthers *et al.* (1973) or Suchow *et al.* (1970). In either case, it is assumed that the effect on the lattice parameter of rare-earth octahedral substitution either is known experimentally or is directly related to the mole fraction of rare earth on the octahedral site. Using the latter method, the crystal compositions grown by Brandle *et al.* (1972) can be calculated. These compositions are shown in Table 2 along with the initial liquid compositions. As pointed out by Carruthers *et al.* (1973), the remarkable fact shown in Table 2 is the small range of crystal compositions which were produced for the large change in liquid compositions. These results suggest that there exists a range of solid solubility in the Gd_2O_3–Ga_2O_3 phase diagram which is inaccessible to the growing crystal.

Geller *et al.* (1972), using ceramic samples, deduced that the correct

Table 2 Liquid–solid compositions for GGG.[a]

Mole % Gd (liquid)	Mole % Gd (solid)
39.76	37.99
39.10	37.91
38.65	37.89
38.20	37.85
37.74	37.84
37.50	37.83
37.04	37.82
36.59	37.81
36.14	37.79
35.71	37.76

[a] From Brandle et al. (1972).

composition for "pulled" GGG crystals is $Gd_{3.03}Ga_{4.97}O_{12}$. Allibert et al. (1974) report the congruent composition for "pulled" GGG to be $Gd_{3.05}Ga_{4.95}O_{12}$ in close agreement with that reported by Geller et al. (1972). They also report the formation of a gallium-rich composition, $Gd_{2.98}Ga_{5.02}O_{12}$ having a lattice parameter of 12.372 Å which was not observed by Carruthers et al. (1973).

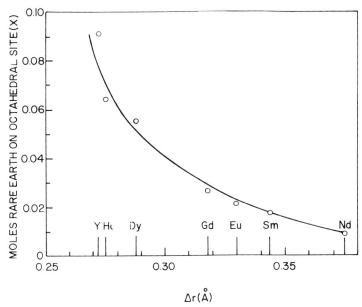

Fig. 10 Ionic radius of the rare-earth ion versus the moles of rare earth substituting on the octahedral site for Czochralski-grown material. (From Brandle and Barns, 1974, with permission of the North Holland Publishing Co.)

Brandle and Barns (1974) have confirmed the congruent composition of GGG and extended those results to other rare-earth gallium garnets. They found that as the ionic radius of the rare earth on the dodecahedral site decreased, the degree of substitution of the rare earth on the octahedral site also increased, i.e. the congruent crystal composition shifted more toward the rare-earth rich region of the phase diagram. Their results are shown graphically in Fig. 10.

Thus, it is fairly well established that Czochralski-grown gadolinium gallium garnet is not stoichiometric, but is gadolinium-rich with an x value of about 0.03; that there exists a wide range of solid solubility of gadolinium in $Gd_3Ga_5O_{12}$ that is not accessible to the growing crystal and very little, if any, gallium solubility exists in $Gd_3Ga_5O_{12}$. Furthermore, variations in stoichiometry which do occur in the crystal have little effect upon the bulk lattice parameter, although variations can be expected across growth striations.

6 OTHER GARNET COMPOSITIONS

The garnet structure first solved by Menzer (1928) consists of an oxygen framework with three different, distinct cation sites. The largest cation, usually a rare earth, occupies the dodecahedral or "c" site, i.e. the cation is surrounded by eight oxygen atoms. The second site is an octahedral or "a" site in which a cation of intermediate size is located and is surrounded by six oxygen atoms. The third site or "d" site is a tetrahedral site and is usually occupied by the smallest cation. Thus, the garnet formula can be written as

$$\{A_3\}[B_2]C_3O_{12}, \tag{2}$$

where the A cations are in dodecahedral coordination, the B cations are in octahedral coordination and the C cations are in tetrahedral coordination. For many of the simple garnets, the same cation is located on both the octahedral (a) site and the tetrahedral (d) site, e.g. $Ln_3Fe_5O_{12}$, $Ln_3Al_5O_{12}$ and $Ln_3Ga_5O_{12}$. As might be expected, because there are three different sites within the garnet structure, selective substitution of various cations into the different sites can be accomplished to alter the magnetic properties of the film (Nielsen, 1971; Van Uitert et al., 1970). A similar approach can be used to change the lattice parameter of the substrate crystal to meet new material requirements which might be necessary to achieve faster operating speeds and smaller bubble diameters.

One of the first requirements for these new films could be a lattice parameter larger than that of GGG. As has already been mentioned, several groups of garnets can meet the larger lattice parameter requirement:

(1) Other rare-earth gallium garnets.
(2) Solid solutions of rare-earth gallium garnets.
(3) Rare-earth scandium aluminum garnets and rare-earth scandium gallium garnets.
(4) Coupled substitution gallium garnets.

All of these classes of garnets meet the first requirement for growth—they all form stable compounds up to their melting points and, therefore, can be grown using the Czochralski growth process. Each class also contains members which have a lattice parameter in the desired range 12.42 Å to 12.51 Å. The potential use of each material will depend upon the difficulties encountered during growth to achieve the desired crystal perfection requirement, which will remain the same regardless of the material.

6.1 Ion substitution

Substitution of different ions on the various crystallographic sites becomes more complex when the substituting ions do not exhibit complete site selectivity. The high growth temperature and ions of similar size can contribute to several different types of substitution, as well as mixed types.

The first type is pure dodecahedral or "c" site substitution, and can be represented by the general formula

$$\{A_{3-x}B_x\}[Ga_2]Ga_3O_{12}, \tag{3}$$

where A and B represent different rare-earth ions. Substitution of this type would be a solid solution of two different rare-earth gallium garnets (Brandle and Valentino, 1972). A second form of dodecahedral substitution can result when the ion on the octahedral (a) site is sufficiently large to occupy some of the positions on the dodecahedral site, and is represented by

$$\{A_{3-x}B_x\}[B_2]Ga_3O_{12}. \tag{4}$$

Suchow et al. (1970) and Suchow and Kokta (1972) have reported on this type of substitution where the B ion is a smaller rare-earth ion.

A second type of substitution can involve the octahedral (a) site and has already been discussed for the rare-earth gallium garnets. This type can be represented by

$$\{A_3\}[A_xGa_{2-x}]Ga_3O_{12} \tag{5}$$

and, as can be seen, involves the direct substitution of the rare-earth ion into the octahedral lattice site. GGG and the other rare-earth gallium garnets are a special case of this type of substitution. An alternate form of octahedral substitution can involve the tetrahedral (d) site ion and is represented by

$$\{A_3\}[B_{2-x}Al_x]Al_3O_{12}. \tag{6}$$

This form of substitution has been proposed by Chow *et al.* (1974) for some scandium-containing garnets.

A third type of garnet composition involves the coupled substitution of two different ions on the dodecahedral and octahedral sites of the garnet structure, and can be represented by the general formula

$$\{A_{3-x}A'_x\}[B_{2-y}B'_y]Ga_3O_{12} \tag{7}$$

For the case where A is a rare-earth ion and B is a gallium ion, the valence sum of A' plus B' must be six, and x must equal y.

As can be seen from the above discussion, as the crystal composition becomes more complex, the possible types of ion substitution increases and it is very probable that several types of substitution coexist at the growth temperature.

6.2 Gallium garnets

Within the class of rare-earth gallium garnets, there exist several additional garnets which could be used as an alternative substrate material with lattice parameters larger than GGG and are shown in Table 3. All have been grown by the Czochralski technique (Brandle and Valentino, 1972) and processed into substrates for use in LPE growth. Of those listed in Table 3, only two, $Nd_3Ga_5O_{12}$ and $Sm_3Ga_5O_{12}$, can be considered as an alternative substrate material. The third, $Eu_3Ga_5O_{12}$, can be eliminated because of its high cost.

For $Nd_3Ga_5O_{12}$ and $Sm_3Ga_5O_{12}$, growth becomes more difficult even though the melting points are lower, 1515°C and 1620°C respectively (Brandle and Valentino, 1972). The major problem in their growth is associated with interface stability. Obtaining the proper thermal geometry

Table 3 Comparison of various gallium garnet substrates.

Substrate	a_0[a]	Cost[b]	Growth rate[b]
$Gd_3Ga_5O_{12}$	12.3831	1.00	1.0
$Eu_3Ga_5O_{12}$	12.4084	5.37	0.9
$Sm_3Ga_5O_{12}$	12.4383	0.90	0.8
$Nd_3Ga_5O_{12}$	12.5090	0.87	0.6

[a] From Brandle and Barns (1974).
[b] Relative to $Gd_3Ga_5O_{12}$.

to prevent the interface from becoming concave during growth, thereby leading to impurity entrapment and dislocations, is extremely difficult. Also, because of a lower thermal conductivity, the crystals have a tendency to show more pronounced faceting resulting in larger strain areas in the crystal cross section.

Despite these problems, both $Nd_3Ga_5O_{12}$ and $Sm_3Ga_5O_{12}$ have been grown with good internal quality sufficient for use as an LPE substrate. These crystals have been typically one inch or less in diameter. The growth of these materials in a size comparable to that of GGG (3 to 4 in. diameter) would require substantial work on furnace design and growth technique to achieve the quality now possible in GGG.

6.3 Solid solutions

Another class of compounds examined for alternate substrate materials was that of solid solutions between various rare-earth gallium garnets, and involves only dodecahedral ion substitution. Varnerin (1971) has pointed out many of the advantages of the use of such solid solutions, among them, the accessibility of a continuous range of lattice parameters from which to choose, rather than the discrete "steps" one finds when going from one rare-earth garnet to another. Heinz et al. (1972b) also found advantages in the use of rare-earth gallium garnet solid solutions. Brandle and Valentino (1972) investigated several gallium garnet solid systems to determine their distribution coefficients and growth parameters.

Regardless of the advantages, these crystals are solid solutions and hence, their properties are dependent upon the liquid composition. The major drawback to the use of solid-solution gallium garnet crystals comes from the lack of precise lattice parameter control through the length of the crystal. This is illustrated by using the data of Brandle and Valentino (1972) in Table 4 for the distribution coefficients in the $(Nd, Sm)_3Ga_5O_{12}$ and the $(Sm, Gd)_3Ga_5O_{12}$ system.

If one assumes a maximum tolerable lattice parameter variation of ± 0.001 Å in the substrate, then the crystal variation from top to bottom can be no greater than 0.002 Å. Table 4 illustrates the variations one can expect for various gallium garnet solid solution systems. As can be seen in the table, the variation will exceed the limit of 0.002 Å when 50% of the liquid has been crystallized in the $(Sm, Gd)_3Ga_5O_{12}$ system, and when only about 30% of the liquid has been crystallized in the $(Nd, Sm)_3Ga_5O_{12}$ system. Thus, the use of solid solutions severely limits the usable amount of initial material.

In addition to lattice parameter variations through the length of the

Table 4 Lattice parameter variation for solid solutions.[a]

Solid solution system	Fraction of liquid crystallized		Δa_0
	0.01	0.50	
25% Gd–75% Sm	12.4215	12.4236	0.0021
50% Gd–50% Sm	12.4077	12.4097	0.0020
75% Gd–25% Sm	12.3934	12.3958	0.0024
25% Sm–75% Nd	12.4868	12.4903	0.0035
50% Sm–50% Nd	12.4689	12.4723	0.0034
75% Sm–25% Nd	12.4515	12.4545	0.0030

[a] Distribution coefficients for calculations from Brandle and Valentino (1972).

crystal, other problems associated with growth severely limit the usability of the crystals. Since these are gallium garnets, all the problems associated with pure gallium garnets as discussed earlier are present, but to a greater extent because of the solid solution nature of the crystals. Thus, control of crystal perfection becomes more difficult, particularly inclusions and dislocations; growth rates must be reduced to avoid segregation effects and atmosphere control is more critical. Because of these problems in growth and the problem of lattice parameter variations in the crystal, solid solutions of the gallium garnets cannot be considered as viable substrate materials for future devices and alternate material compositions must be considered.

6.4 Scandium substituted garnets

A second class of garnets which has several members with lattice parameters suitable for substrate use is the group of scandium substituted garnets, as shown in Fig. 11. Several investigators (Geller, 1967) have reported the selective substitution of scandium into the octahedral site of various irons containing rare-earth garnets. Brandle and Barnes (1973) and Chow et al. (1974) have reported on the Czochralski growth of scandium-substituted garnets. Two types of substituted garnets are possible, with the type depending on whether gallium or aluminium is on the tetrahedral site. For the scandium gallium garnets, $Y_3Sc_2Ga_3O_{12}$ has a lattice parameter in the desired range and has been grown using the Czochralski technique. As in the case for the pure gallium garnets, the crystal composition is not stoichiometric but contains ion substitutions similar to those of GGG. Brandle and Barns (1973) have proposed that the crystals contain excess yttrium substituting for scandium on the octahedral site, and can be represented by the general formula $Ln_3[Ln_xSc_{2-x}]Ga_3O_{12}$. Their results are

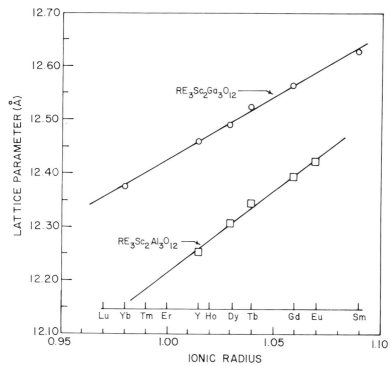

Fig. 11 Lattice parameters of Czochralski-grown rare-earth scandium gallium and aluminum garnets. (From Brandle and Barns, 1973, with permission of the North Holland Publishing Co.)

based on the assumption that the observed lattice parameter of the crystal is larger than that calculated for stoichiometric $Y_3Sc_2Ga_5O_{12}$. Chow et al. (1974) have proposed that the composition is gallium rich and can be represented by $Ln_3[Sc_{2-x}Ga_x]Ga_3O_{12}$, i.e. gallium substituting for scandium on the octahedral site. Their results are based on lattice parameter measurements indicating that the observed lattice parameter of the crystal is smaller than the stoichiometric compound.

Because of the close size of the ions involved, the probable correct composition involves more than one type of substitution. The exact composition can only be determined by obtaining accurate lattice parameter measurements for the stoichiometric materials and crystals.

Regardless of the exact composition, it has been shown that these compounds can be grown by the Czochralski technique. Because they are gallium garnets, they retain many of the growth properties of the pure gallium garnets, and, therefore, have many of the same problems. Composition control is more difficult because of higher melting points (Ga_2O_3 loss), and one is now working in a ternary rather than a binary system. Aside from this problem, these crystal compositions can be grown with high internal perfection using growth conditions similar to those of the rare-earth gallium garnets. They, therefore, provide a viable class of compounds from which to choose alternate substrate compositions.

The second type of scandium-substituted garnet which can serve as a substrate material involves the use of aluminum on the tetrahedral site of which $Gd_3Sc_2Al_3O_{12}$ is a typical example. In these crystals it has been reported (Brandle and Barns, 1973) that the lattice parameter of the crystal is smaller than that of the stoichiometric compound, again indicating a nonstoichiometric congruent composition. Two forms of substitution can occur which can explain this observed difference: (1) scandium substituting on the dodecahedral site, or (2) aluminum substituting on the octahedral site, i.e. $\{Ln_{3-x}Sc_x\}Sc_2Al_3O_{12}$ or $Ln_3[Sc_{2-x}Al_x]Al_3O_{12}$ respectively. Further work on the gadolinium compound has shown that both types of substitution occur simultaneously, and that the congruent composition is $\{Gd_{2.91}Sc_{0.09}\}[Sc_{1.75}Al_{0.25}]Al_3O_{12}$. Such large degrees of site substitution make crystal growth more difficult because of strain and worse still, can drastically alter the lattice parameter of a crystal. As an example, a crystal grown from a stoichiometric melt of $Gd_3Sc_2Al_3O_{12}$ has a lattice parameter of 12.395 Å (Brandle and Barns, 1973) whereas the lattice parameter of a crystal grown from the congruent composition is 12.378 Å (C. D. Brandle, 1974 unpublished observations). Furthermore, for this class of materials, the degree of site substitution is dependent upon the compound which makes the lattice parameter and the congruent composition difficult to predict.

A second problem which is also the result of site substitution is growth striations and the associated strain. This is much more severe in the scandium aluminum garnets than in the scandium gallium garnets because of the wider range of site substitution available to the crystal. This strain can be of sufficient magnitude to result in cracking of the crystal when it is cooled. Also, because of a higher thermal conductivity in the scandium aluminum garnets than in the scandium gallium garnets, it is more difficult to grow the aluminum garnets with a flat interface. Although the rare-earth scandium aluminum garnet class has several compounds which have lattice parameters in the desired range, the use of these materials at present is restricted because material of adequate quality cannot be produced.

6.5 Coupled substitution

As work continues on new magnetic film compositions for bubble devices and optical applications, new garnets with larger lattice parameters become

Table 5 Summary of various garnet substrate materials.

Substrate material	Advantages	Disadvantages
$Ln_3Ga_5O_{12}$	Cost comparable to GGG Suitable lattice parameter Can be grown strain free	Comparable quality to GGG more difficult to obtain Slower growth rates required
$\{Ln_{3-x}Ln_x\}Ga_5O_{12}$	Adjustable lattice parameter Cost comparable to GGG Can be grown strain-free	Lattice constant varies through length of crystal Slower growth rates required Quality difficult to maintain
$Ln_3Sc_2Ga_3O_{12}$	Suitable lattice parameter Growth rate similar to GGG Can be grown strain-free	Cost about twice that of GGG Some solid solution exists Higher melting points
$Ln_3Sc_2Al_3O_{12}$	Suitable lattice parameter Little vapor loss	Slower growth rates required Costs about twice that of GGG Difficult to grow strain-free Considerable solid solution on octahedral and dodecahedral sites
$\{Ln_{3-x}A_x\}[Ga_{2-x}B_x]Ga_3O_{12}$ or $Ln_3[Ga_{2-2x}A_xB_x]Ga_3O_{12}$	Suitable lattice parameter Cost comparable to GGG Can be grown strain-free	Distribution coefficient function of liquid composition Slower growth rates required

necessary. The Czochralski growth of a new class of garnets has been reported by Mateika et al. (1975) which involves the coupled substitution of two cations into the dodecahedral and octahedral sites of the garnet structure.

As with the other forms of site substitution previously discussed, this type also presents problems in lattice parameter control. The distribution coefficient for the substituting A' and B' ions as shown in Eq. (7) is not necessarily unity, and has been found to vary with the value of x (Mateika et al., 1975). Despite this problem, there exists a value x for which the distribution coefficient is unity, and crystals having the composition $\{Gd_{3-x}Ca_x\}[Ga_{2-y-z}Zr_yGd_z]Ga_3O_{12}$ with $x = y = 0.45$ and $z = 0.04$ have been grown with a constant parameter (Mateika et al., 1975).

This work has been extended to include Mg^{2+} and Zr^{4+} substitutions on the octahedral site of GGG to form crystals having a composition of $Gd_3[Mg_xZr_xGd_zGa_{2-2x-z}]Ga_3O_{12}$ (Mateika and Rusche, 1977). However, for this type of substitution, distribution coefficients of unity for Zr^{4+} could not be achieved. Thus, these crystals showed a gradual shift in lattice parameter throughout their length. Recent work (Mateika et al., 1982) using a similar type of substitution in other rare-earth gallium garnets has shown that this type of substitution can provide garnet crystals with expanded lattice parameters, but, as mentioned above, the distribution coefficients are not unity. In addition to these substituted gallium garnets, additional complexity has been achieved (Mateika et al., 1982) by using more than one divalent ion to charge compensate Zr^{4+}. Crystals of $Gd_{3-x}Ca_xGa_{5-x-2y}Mg_yZr_{x+y}O_{12}$ have been grown with low dislocation densities and up to 82 mm in diameter.

Of the several garnet compositions discussed only two types have proved useful as a substrate material. The first type is the rare-earth gallium garnets of which GGG ($Gd_3Ga_5O_{12}$) is the substrate of choice. Other rare-earth gallium garnets such as $Nd_3Ga_5O_{12}$ and $Sm_3Ga_5O_{12}$ have been used but only for experimental films. The second type which has found limited use is the coupled substitution garnets of GGG.

All others, either because of lattice parameter considerations and/or growth considerations have been eliminated from further consideration as substrates for magnetic bubbles. Table 5 gives a summary of the various types of possible garnet substrate materials along with their advantages and disadvantages.

REFERENCES

Allibert, M., Chatillon, C., Mareschal, J., and Lissalde, F. (1974). *J. Cryst. Growth* **23**, 289.
Basterfield, J., Prescott, M. J., and Cockayne, B. (1968). *J. Mater. Sci.* **3**, 33.
Belt, R. F., and Moss, J. P. (1973). *Mater. Res. Bull.* **8**, 1197.
Belt, R. F., Moss, J. P., and Latore, J. R. (1973). *Mater. Res. Bull.* **8**, 357.
Blank, S. L., and Nielsen, J. W. (1972). *J. Cryst. Growth* **17**, 302.
Blank, S. L., Hewitt, B. S., Shick, L. K., and Nielsen, J. W. (1973). *AIP Conf. Proc.* No. **10**, 256.
Brandle, C. D. (1977). *J. Cryst. Growth* **42**, 400.
Brandle, C. D. (1978). *J. Appl. Phys.* **49**, 1855.
Brandle, C. D. (1982). *J. Cryst. Growth* **57**, 65.

Brandle, C. D., and Barns, R. L. (1973). *J. Cryst. Growth* **20**, 1.
Brandle, C. D., and Barns, R. L. (1974). *J. Cryst. Growth* **26**, 169.
Brandle, C. D., and Valentino, A. J. (1972). *J. Cryst. Growth* **12**, 2.
Brandle, C. D., Miller, D. C. and Nielsen, J. W. (1972). *J. Cryst. Growth* **12**, 195.
Carruthers, J. R. (1976). *J. Cryst. Growth* **36**, 212.
Carruthers, J. R., Kokta, M., Barnes, R. L., and Grasso, M. (1973). *J. Cryst. Growth* **19**, 204.
Chow, K., Keig, G. A., and Hawley, A. M. (1974). *J. Cryst. Growth* **23**, 58.
Cockayne, B., and Roslington, J. M. (1973). *J. Mater. Sci.* **8**, 601.
Cockayne, B., Chesswas, M., and Gasson, D. B. (1968). *J. Mater. Sci.* **3**, 224.
Cockayne, B., Roslington, J. M., and Vere, A. W. (1973). *J. Mater. Sci.* **8**, 382.
Dash, W. C. (1959). *J. Appl. Phys.* **30**, 459.
Geller, S. (1967). *Z. Kristallogr., Kristallgeom., Kristallphys., Kristallchem.* **125**, 1.
Geller, S., Espinosa, G. P., and Crandall, P. B. (1969). *J. Appl. Crystallogr.* **2**, 86.
Geller, S., Espinosa, G. P., Fullmer, L. D., and Crandall, P. B. (1972). *Mater. Res. Bull.* **7**, 1219.
Glass, H. L. (1972). *Mater. Res. Bull.* **7**, 1087.
Glass, H. L., Besser, P. J., and Hamilton, T. N. (1973). *Mater. Res. Bull.* **8**, 309.
Heinz, D. M., Besser, P. J., Owens, J. M., Mee, J. E., and Pulliam, G. R. (1972a). *J. Appl. Phys.* **42**, 1243.
Heinz, D. M., Moudy, L. A., Elkins, P. E., and Klien, D. J. (1972). *J. Electron. Mater.* **1**, 310.
Keig, G. A. (1973). *AIP Conf. Proc.* No. **10**, 237.
Kikuta, S., Kohra, K., and Sugita, Y. (1966). *Jpn. J. Appl. Phys.* **5**, 1047.
Kishino, S., Isomae, S., Takagi, K., and Ishii, M. (1974). *Mater. Res. Bull.* **9**, 1301.
Lal, K., and Mader, S. (1976). *J. Cryst. Growth* **3**, 357.
Laudise, R. A. (1970). "The Growth of Single Crystals", Prentice-Hall, Englewood Cliffs, New Jersey.
Levinstein, H. J., Landorf, R. W., and Licht, S. J. (1971). *IEEE Trans. Magn.* **MAG-7**, 470.
Linares, R. C. (1964). *Solid State Commun.* **2**, 229.
Linares, R. C. (1968). *J. Cryst. Growth* **3**, 443.
Mateika, D., and Rusche, C. (1977). *J. Cryst. Growth* **42**, 440.
Mateika, D., Herrnring, J., Rath, R., and Rusche, C. (1975). *J. Cryst. Growth* **30**, 311.
Mateika, D., Lurien, R., and Rusche, C. (1982). *J. Cryst. Growth* **56**, 677.
Matthews, J. W., Klokholm, E., Sadagopan, V., Plaskett, T. S., and Mendel, E. (1973). *Acta Metall.* **21**, 203.
Mee, J. E., Pulliam, G. R., Archer, J. L. and Besser, P. J. (1969). *IEEE Trans. Magn.* **MAG-5**, 289.
Menzer, G. (1928). *Z. Kristallogr., Kristallgeom., Kristallphys., Kristallchem.* **69**, 300.
Metselaar, R., and Huyberts, M. A. H. (1973). *J. Phys. Chem. Solids* **34**, 2257.
Miller, D. C. (1973). *J. Electrochem. Soc.* **120**, 678.
Miller, D. C., and Caruso, R. (1974). *J. Cryst. Growth* **27**, 274.
Miller, D. C., and Pernell, T. L. (1981). *J. Cryst. Growth* **53**, 523.
Miller, D. C., and Pernell, T. L. (1982). *J. Cryst. Growth* **57**, 253.
Miller, D. C., Valentino, A. J., and Shick, L. K. (1978). *J. Cryst. Growth* **44**, 121.
Morizane, K., Witt, A. F., and Gatos, H. C. (1967). *J. Electrochem. Soc.* **114**, 738.
Nes, E. (1974). *Acta Metall.* **22**, 81.
Nielsen, J. W. (1971). *Metall. Trans.* **2**, 625.

O'Kane, D. F., Sadagopan, V., and Giess, E. A. (1973). *J. Electrochem. Soc.* **120**, 1272.
Rósencwaig, A., Tabor, W. J., and Pierce, R. D. (1971). *Phys. Rev. Lett.* **26**, 779.
Schwuttke, G. H. (1962). *In* "Direct Observations of Imperfections in Crystals" (J. B. Newkirk and J. H. Wernick, eds.), p. 497. Wiley (Interscience), New York.
Shannon, R. D., and Prewitt, C. T. (1969). *Acta Crystallogr., Sect. B* **B25**, 925.
Stacy, W. T. (1974). *J. Cryst. Growth* **24/25**, 137.
Stacy, W. T., and Enz, U. (1972). *IEEE Trans. Magn.* **MAG-8**, 268.
Stacy, W. T., Metselaar, P. K., Larson, P. K., Briel, A., and Pistorius, J. A. (1974a). *Appl. Phys. Lett.* **24**, 254.
Stacy, W. T., Pistorius, J. A., and Janssen, M. M. (1974b). *J. Cryst. Growth* **22**, 37.
Suchow, L., and Kokta, M. (1972). *J. Solid State Chem.* **5**, 329.
Suchow, L., Kokta, M., and Flynn, V. J. (1970). *J. Solid State Chem.* **2**, 137.
Takagi, K., Fukazawa, T., and Ishii, M. (1976). *J. Cryst. Growth* **32**, 89.
Tominaga, H., Sakai, M., Fukuda, T., and Namikato, T. (1974). *J. Cryst. Growth* **24/25**, 272.
Van Uitert, L. G., Bonner, W. A., Grodkiewicz, W. H., Pictroski, L., and Zydzik, G. J. (1970). *Mater. Res. Bull.* **5**, 825.
Varnerin, L. J. (1971). *IEEE Trans. Magn.* **MAG-7**, 404.
Whiffin, P. A. C., and Brice, J. C. (1971). *J. Cryst. Growth* **10**, 91.
Witt, A. F., Lichtensteiger, M., and Gatos, H. C. (1973). *J. Electrochem. Soc.* **120**, 1119.
Zupp, R. R., Nielsen, J. W., and Vitorio, P. V. (1969). *J. Cryst. Growth* **5**, 269.

2 Amorphous Gd–Co Alloys for Magnetic Bubble Applications

P. CHAUDHARI, C. H. BAJOREK and M. H. KRYDER

*IBM Thomas J. Watson Research Center
Yorktown Heights, New York*

1	Introduction and general background	31
2	Film fabrication	34
3	Atomic arrangement in amorphous Gd–Co alloys	41
4	Magnetic moment	47
5	Annealing and ion implantation	55
6	Magnetic anisotropy	59
7	Domain wall dynamics of amorphous films	66
8	Amorphous film bubble devices	75
	8.1 Device performance	78
	8.2 Dielectric breakdown	81
9	Summary and conclusions	85
	References	86

1 INTRODUCTION AND GENERAL BACKGROUND

In this paper we shall survey the properties of amorphous alloys for magnetic bubble applications. Since the subject of magnetism in amorphous alloys is relatively new, we believe it might be useful to provide a general description of the structure and properties of amorphous materials before proceeding to the main subject matter of this chapter.

It is commonly accepted that a characteristic property of the atomic structure of amorphous materials is the absence of long-range translational correlations in the equilibrium positions of the constituent atoms. Amorphous materials have, however, considerable short-range order or correlations. By this we mean that the number and positions of the near neighbor atoms is not random. It is possible to describe the average atomic structure of amorphous metallic alloys using the hard sphere models introduced by Bernal and associates for liquids (Chaudhari and Turnbull, 1978). More sophisticated models using soft potentials rather than hard sphere potentials are currently used to obtain better agreement between experiment and model calculations. In general the experimental data suggests that in a monoatomic amorphous metallic solid an atom has on the average 12 nearest neighbors. These atoms are located at distances which are close to the values found in their crystalline counterparts. In multicomponent amorphous alloys the number of near neighbor atoms may not be 12 and considerable deviations from the crystalline values of near neighbor distances may also occur. We shall return to this problem when we discuss the atomic structure of Gd–Co alloys in a later part of this chapter.

A characteristic property of amorphous metallic alloys is their high electrical resistivity and the small temperature coefficient of resistivity. In crystalline alloys electrical resistivity generally decreases as the temperature decreases. This is attributed to a decreased scattering of the charge carriers by phonons. The mean free path of electrons varies with temperature and at low temperature can be several thousands of Angstroms. In contrast to this behavior, positional disorder and/or compositional scattering dominantly determine the mean free path of electrons in an amorphous alloy. This scattering mechanism is relatively insensitive to temperature and hence the small temperature coefficient of resistivity and the relatively large value of resistivity in the amorphous form. The mean free path for electrons (or holes) in amorphous metallic solids is estimated to be of the order of 10 Å.

This short mean free path has several consequences on magnetic phenomena. For example, the extraordinary Hall coefficients are found to be a factor of 100 larger than in the corresponding crystalline phases (Lin, 1969; Okamoto et al., 1976; McGuire et al., 1977a). These rather large values of the extraordinary Hall coefficients are frequently used to characterize amorphous magnetic bubble materials. The Hall voltage is used to determine the saturation moment and the temperature dependence of magnetization. It is conceivable that these large extraordinary Hall coefficients will find applications as sensors for magnetic fields. The short mean free path of electrons is also expected to influence the RKKY interaction (de Gennes, 1962). The spatial range over which we expect the conduction electrons to maintain their polarization is determined by scattering. In an amorphous alloy we expect both the range and amplitude of the electron polarization to be decreased.

In the rare-earth materials where the moment per atom is associated with the deep inner 4f shells, we do not expect a change in moment per atom due to a change in structure. This is generally borne out by experiments. In the case of 3d transition metals such as Ni, Fe, and Co where a band model for magnetism may be more valid, we find that a change in structure or composition affects the moment (Mizoguchi, 1976). Actually, we expect that the 3d metals and alloys might be sensitive to structure or composition even if a local picture is applicable provided the d band, or the bonding associated with this band, is perturbed. Experimentally, the moment of nickel and iron is found to change whereas that of cobalt is unaltered on going from the crystalline to the amorphous form. In the case of iron both the coordination number and nearest neighbor distance in the amorphous form resembles fcc iron rather than bcc iron.

Exchange interactions between magnetic atoms result in ferromagnetic or ferrimagnetic alignment. In an amorphous solid, structural and compositional fluctuations are likely to be enhanced over crystalline values. The effect of these fluctuations is to smear out the value of internal fields over some mean value. However, the directions of such fields or their gradients can have all possible orientations unless the amorphous solid has built into it some structural anisotropy. (From a practical standpoint such structural anisotropy is invariably present due to fabrication procedures and is discussed in some detail later in this chapter.) If the exchange field is sensitive to orientation, we can expect that a change in structure results in a dramatic change in the magnetic alignment. Garnets are a good example of this class. Ferrimagnetic garnets when deposited in the amorphous form show little ferromagnetic response at room temperature (Cuomo et al., 1972). In garnets the exchange interaction is mediated by the oxygen which requires that it be present at the correct bonding angles. Once this angle is perturbed, as it is expected to be in the amorphous form, the interaction is weakened.

The effect of compositional or structural fluctuations is, as indicated before, to introduce a range of exchange interaction values. This can be observed in Mossbauer (Sharon and Tsuei, 1972; Heiman and Lee, 1975a; Forester, 1976) and nuclear magnetic resonance (Raj et al., 1976; Durand, 1976) studies where the hyperfine fields are measured and found to have relatively broad distributions in comparison to the discrete crystalline values. Fluctuations also lead to departures from the Brillouin function fit to magnetization data at high temperatures (Tsuei and Lillienthal, 1976). At temperatures close to absolute zero the familiar $T^{3/2}$ law characteristic of spin wave excitations has been observed in amorphous magnetic alloys (Cochrane and Cargill, 1974). Spin wave modes have also been noted in neutron scattering and ferromagnetic resonance experiments (Mizoguchi, 1976; Axe et al., 1975; McGoll et al., 1976).

Amorphous magnetic alloys of the 3d transition metals show coercivities ranging from a few millioersted to a few oersteds. Amorphous alloys containing rare-earths can, however, show large coercivities. This is generally associated with large fluctuations in the local anisotropy due to variations in the direction of the electrostatic field gradient. We shall return to this point later in this paper. The combination of low coercivity and high electrical resistivity suggests application in the transformer industry.

One of the advantages of amorphous materials is that the range of compositions which can be prepared is not restricted by the equilibrium phase diagrams. A corollary of this attribute is that amorphous alloys are not thermodynamically stable. They transform at some finite temperature range to their crystalline counterpart. Annealing in amorphous alloys prior to crystallization has also been observed. This phenomenon is poorly understood but has been used to advantage in controlling magnetic anisotropy in some transition metal–metalloid systems. In the rare-earth transition metal alloys such as those used in magnetic bubble applications, this has been problematic and has limited the temperature range of device processing and the kinds of compositions that can be considered for device applications. We shall return to this point later.

In the remaining part of this paper we describe the fabrication, structure, and magnetic properties of rare-earth transition metal alloys. In particular we shall review the properties of Gd–Co alloys and refer to other rare earths only as points of departure. Specifically we shall discuss the fabrication, structure and composition, static and dynamic magnetic properties, ion implantation, and annealing of Gd–Co alloys. We shall then examine our results on device design and performance and finally conclude with a discussion of insulator fabrication and the limitations introduced by this insulator.

2 FILM FABRICATION

Sputtering and evaporation techniques have been used to prepare amorphous films with bubble domain characteristics. Historically, sputtering has been investigated in greater detail and, at present, is the only process that has been used to make bubble device films. Evaporation has emphasized exploratory studies because it offers advantages for rapidly varying film composition over broad ranges.

Film properties are strongly dependent on the deposition process. For example, during sputtering substantial amounts of inert gas can be incorporated in the films. This has been found to dilute the magnetization of rare earth–transition metal systems, an effect which has been found useful for adjusting the absolute value of magnetization and its temperature dependence (Cuomo and Gambino, 1977). It has also been established that materials exhibiting perpendicular anisotropy when deposited by sputtering (Gd–Co) may exhibit planar anisotropy when deposited by evaporation in conventional and ultra-high vacuum systems. However, controlled amounts of oxygen in the vacuum system can induce perpendicular anisotropy (Brunsch and Schneider, 1977; Schneider and Brunsch, 1977). These differences are not fully understood. At present it is not clear which of the two deposition processes will ultimately prevail. From the standpoint of this review we shall concentrate on the sputtering process, if for no other reason than the simple observation that this process is capable of producing device-quality films.

In sputtering a plasma discharge is sustained in an inert gas such as argon. The ionized argon atoms are accelerated by an applied field of several kilovolts towards a "target". This bombardment knocks or sputters off atoms of the target material and these deposit on the substrate, which is usually parallel to the plane of the target. The plasma is sustained either by a dc (dc sputtering) or an rf (rf sputtering) field. The advantage of dc sputtering lies in the simplicity of the electronics. Its disadvantage lies in the transient layer that is formed on an electrically insulating substrate. In order to avoid this transient layer most of the bubble materials are deposited by rf sputtering (Cuomo and Gambino, 1975; Bajorek and Kobliska, 1976). The important features of the sputtering process and its control consist of stable electronics, stable inert gas pressure, and temperature control of substrate and target. External but related to sputtering are target composition and uniformity, sputtering geometry, and substrate holder. In general, film composition depends on target composition and deposition parameters. Sputtering instrumentation designed to apply power to both target and substrate electrodes, sketched in Fig. 1a, has been found to be useful in achieving the degree of film composition control required by practical applications. The relative power levels applied to the two electrodes are customarily monitored by the ratio of dc potentials, either directly applied in dc sputtering or spontaneously developed in rf sputtering, at the two electrodes (Vossen, 1971).

Sputtering and resputtering rates for the constituents of alloys are generally unequal. The film composition is therefore not equal to that of the target composition. However, there is a well defined relation between the two. The correlation between film and target composition is dependent on geometric parameters such as target size and interelectrode separation, and deposition factors such as gas pressure. These aspects are qualitatively sketched in Figs 1b and 1c for an amorphous Gd–Co–Mo alloy.

Figure 1b represents the Co/(Co + Gd) atomic ratio versus substrate/target voltage ratio with constant sputtering gas pressure and system geometry. At zero substrate voltage (power), the film composition is approximately equal to the target composition. At higher substrate voltages, films are rich in Co

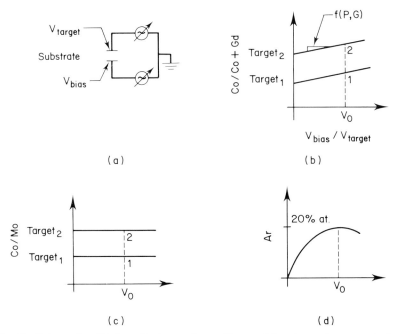

Fig. 1 Schematic of (a) sputtering system with provision to apply power to both target and substrate electrodes, (b) Co/(Co + Gd) atomic ratio versus substrate/target voltage ratio V_0, with constant sputtering gas pressure and system geometry, (c) Co/Mo atomic ratio versus V_0, and (d) Ar content in the sputtered film versus V_0.

relative to the target due to preferential loss of Gd to the grounded surfaces of the sputtering chamber. The slopes of the curves in Fig. 1b depend on gas pressure and geometry and are smallest for high pressures and small interelectrode separation to target diameter ratios. By contrast, as indicated in Fig. 1c, the Co/Mo ratio in the film is much less dependent upon deposition parameter. This is expected, as Co and Mo have nearly equal sputtering and resputtering rates as well as nearly equal escape rates to the grounded surfaces of a vacuum chamber.

Film growth rate and uniformity are also dependent on gas pressure, rf power, and system geometry (Cuomo and Gambino, 1975; Bajorek and Kobliska, 1976; Vossen, 1971). The deposition rate is proportional to pressure and power. Rates of 100 Å/min can be easily achieved with 30 μm Ar pressures and power densities of a few watts per square centimeter of target. Uniformity is dependent on target size and substrate motion. Planetary type substrate motion and target diameters in the vicinity of 20 cm have enabled achieving better than 5% uniformity of all physical and magnetic properties over substrates subtending a substrate holder area of 75% of the target area (Kobliska et al., 1975). Straightforward engineering considerations suggest that much better than 5% uniformity can be achieved over these large dimensions with better system design.

Film composition design and control also require knowledge about the gas content in sputtered films. As shown in Fig. 1d, film gas content is a strong function of substrate voltage (substrate electrode power). It is negligible at zero substrate voltage, reaches a maximum at moderate substrate voltages, and decreases for large voltages (Cuomo and Gambino, 1977). This dependence of gas content versus substrate voltage is typical for films prepared by conventional dc and rf sputtering techniques. It depends primarily on substrate voltage. To first order it is independent of gas pressure and system geometry. The precise mechanism responsible for gas inclusion is not well understood. It is generally ascribed to implantation of Ar ions via the accelerating potential between the neutral plasma and substrate surface (approximately the substrate voltage), gas entrapment under material arriving from the target, and film density related to film deposition and resputtering rates.

The nominal film composition and uniformity of magnetic properties are determined by the target. A considerable amount of effort has therefore been spent in fabricating targets that are reproducible and uniform. The first series of targets that were examined were made from arc-melted materials (Cuomo and Gambino, 1975; Bajorek and Kobliska, 1976). These targets possess a considerable degree of composition variations across the target area. In order to minimize the uniformity variations in the deposited films, the substrates and target holder were moved relative to each other. The arc-melted targets were supplanted by hot-pressed targets, which yielded films whose properties were indistinguishable from those achieved in the arc-melted targets.

Subsequent evaluations have confirmed these earlier results. However, these also identified the need for special precautions when using hot-pressed targets. Depending on preparation and handling methods, the hot-pressed targets, which are not 100% dense, have been found to "absorb" substantial amounts of gases (air, water vapor, etc.). Achieving reproducible film properties and negligible contamination required careful preconditioning of hot-pressed targets by sputtering under accelerated conditions prior to normal use, minimizing or avoiding exposure to uncontrolled atmospheres between depositions, and sorting targets in vacuum, all tractable but undesirable constraints.

There are now two additional target options which avoid some of the problems of the hot-pressed and arc-melted targets. The first consists of using "pie-shaped" segments of foils, as depicted in Fig. 2, in conjunction with circumferential target or substrate rotation. This method is nearly ideal

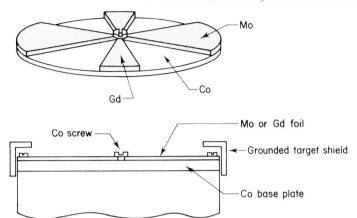

Fig. 2 Segmented target for depositing Gd–Co–Mo alloy films.

for rapid changes of target (film) composition by changing the relative size of the segments with "tin snips" when materials are available in foil form. Figure 3 exemplifies the range of film and target composition that can be rapidly evaluated using such targets. Figure 4 compares some of the properties of films prepared from segmented and arc-melted targets. In brief, segmented targets of this type have been found to be clearly superior to any other target investigated in this laboratory in terms of better flexibility, lower cost, and easier use, especially when used in an exploratory mode.

The second novel target arrangement consists of using a sputtering system

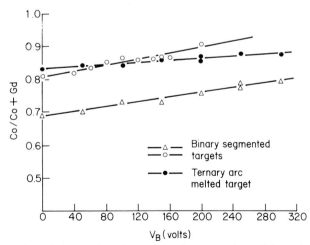

Fig. 3 Co/(Co + Gd) atomic ratio versus substrate voltage (bias voltage) V_B with target voltage held constant.

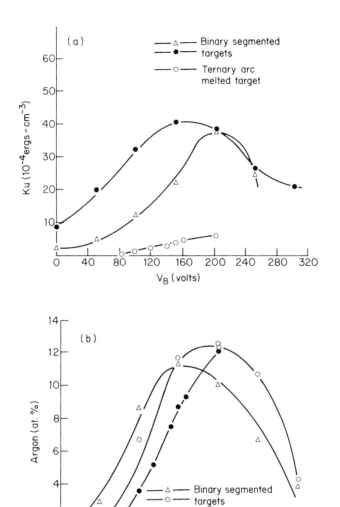

Fig. 4 Comparison of some properties of films made from segmented and arc-melted targets: (a) perpendicular anisotropy K_u versus substrate voltage V_B and (b) argon content in the film versus V_B.

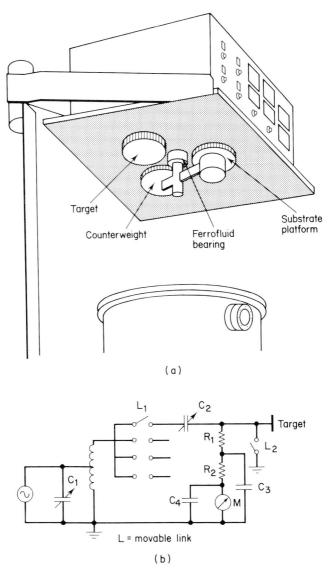

Fig. 5 (a) Sketch showing the basic elements of the multiple-target sputtering system (for clarity, the shutters and two of the targets are omitted from the drawing) and (b) matching network used with multiple-target sputtering system. (Only one target is shown; the circuit consisting of C_2, C_3, C_4, R_1, R_2, and M is duplicated for each target and the substrate platform; unused targets are grounded using link L_2.)

specially modified to independently provide rf power (target voltages) to each of three single element targets (Burilla *et al.*, 1976) (Fig. 5). This feature and rotation of a substrate holder—such that it moves sequentially and periodically under each target, rotation in an orbital mode—allow for varying the composition of the films over broad ranges. This method is expected to be the most flexible of all the techniques that have been examined so far.

3 ATOMIC ARRANGEMENT IN AMORPHOUS Gd–Co ALLOYS

The sputtered and evaporated films of rare-earth transition metal alloys are generally concluded to be amorphous on the basis of x-ray, electron, and neutron diffraction experiments. In this section we review what we know about atomic arrangement in these alloys. Our interest is not just to use this information to calculate magnetic moments but also to deduce the structural mechanism responsible for the origin of magnetic anisotropy. We shall confine our attention to the structure of Gd–Co alloys as these materials have probably been investigated more than any other rare earth–transition metal alloys and also because they come close to satisfying bubble device requirements.

Atomic arrangements in amorphous Gd–Co alloys have been deduced from electron and x-ray diffraction measurements (Cargill, 1974; Wagner *et al.*, 1976; Graczyk, 1976; Nandra and Grundy, 1977). A function that is frequently used to describe atomic arrangements in amorphous solids and which can be determined by Fourier inversion of diffraction data is the radial distribution function. This function gives the average number of atoms between distance r and $r + dr$ from another atom in the solid. The averaging is done over all atoms that are contributing to the scattering of the incident electrons, neutrons, or x-rays. The average number of atoms is $4\pi r^2 \rho(r)\, dr$, where $\rho(r)$ is the density of atoms at distance r, and the radial distribution function is $4\pi r^2 \rho(r)$. The density function $\rho(r)$ approaches the average density ρ_0 of the solid at large r. Typical, experimentally determined, radial distribution functions for amorphous Gd–Co and Gd–Fe alloys are shown in Fig. 6. The position of the first peak in the radial distribution function is the most probable distance between nearest neighbor atoms, and the area under the peak is the co-ordination number corresponding to the peak, i.e. the average number of nearest neighbor atoms. In the case of Gd_{18}–Co_{82}, the first broad maximum appears to consist of three distinct contributions. The first nearest neighbor contribution comes from cobalt atoms at a distance of 2.47 ± 0.05 Å. The number of Co atoms surrounding a cobalt is

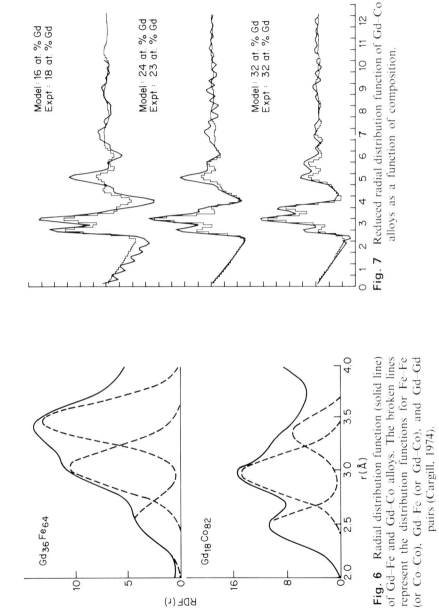

Fig. 6 Radial distribution function (solid line) of Gd–Fe and Gd–Co alloys. The broken lines represent the distribution functions for Fe–Fe (or Co–Co), Gd–Fe (or Gd–Co), and Gd–Gd pairs (Cargill, 1974).

Fig. 7 Reduced radial distribution function of Gd–Co alloys as a function of composition.

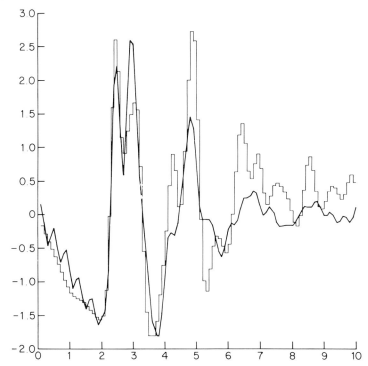

Fig. 8 Reduced radial distribution function of a $Gd_{18}Co_{72}$ alloy. The histogram is a calculated function using a 51-atom crystallite of $GdCo_5$.

on an average 7.2 ± 0.7. The second distance is between Gd and Co atoms at 2.97 ± 0.05 Å, with a Gd–Co co-ordination number of 12 ± 1. The third distance is associated with Gd–Gd pairs at 3.4 Å having a co-ordination number of 3 ± 1. The change in nearest neighbor distance and co-ordination number with composition is shown in Fig. 7.

A function that is frequently used in comparing experimental results and model calculations is the reduced radial distribution function, $G(r)$. This function is related to the radial distribution function through the density function $\rho(r)$ and is given by $G(r) = 4\pi r[\rho(r) - \rho_0]$. As the composition of the amorphous Gd–Co alloy is close to $GdCo_5$ we show the calculated reduced density function for this compound superimposed on the experimental data in Fig. 8. We note that the experimental data and the model calculations show some similarity but do not agree in relative peak heights.

The structure of rare-earth transition metal alloys has also been computer-simulated using a dense random packed hard-sphere model. There have been two approaches to modeling. In one the structure is simulated by starting with a seed tetrahedron and subsequently adding atoms to this seed. The

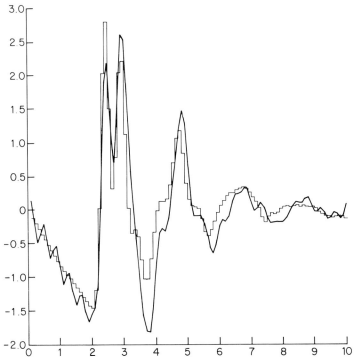

Fig. 9 Reduced radial distribution function of a $Gd_{18}Co_{72}$ alloy. The histogram is a calculated function using 800-atom dense random-packed model.

global model is relaxed with a Lennard–Jones potential using the tetrahedra perfection as a criteria for relaxation (Cargill and Kirkpatrick, 1976). The resultant reduced radial distribution function from such a model is shown and compared with the experimental data in Fig. 9. It is apparent that this model gives a better fit to the data than the microcrystalline $GdCo_5$ structure.

In the second set of simulations the structure is modelled by computer-simulating the deposition of a film by evaporation and sputtering (Kim *et al.*, 1977). Film deposition is simulated by depositing atoms from a prescribed direction onto a substrate. A hard sphere potential is assumed, and the oncoming atoms make contact and move locally until three-point contact is made. In simulations of sputtering the incident stream of atoms has a cosine distribution and the atoms are allowed to bounce from the film with a preassigned probability. The latter simulates bias sputtering. Films deposited by this simulation process are amorphous and an example of the reduced intensity function obtained from such a simulation is shown in Fig. 10. One of the findings in computer simulation of film deposition is the built-in structural anisotropy. It is found that in simulations of evaporation the

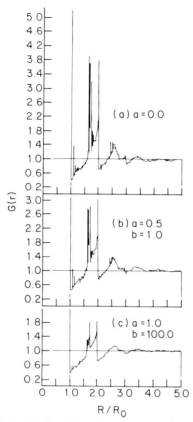

Fig. 10 Reduced intensity function of a computer simulated as-deposited film.

structural anisotropy follows the angle of incidence of vapor stream in remarkable agreement with experimental data. In simulations of sputtering, structural anisotropy is also observed but is not as strong as in the evaporated film. It is reduced both by the cosine distribution of the incident stream of atoms and by the bouncing of atoms.

The structure of amorphous Gd–Co alloys has also been investigated by transmission electron microscopy and electron microdiffraction. Transmission electron microscopy (Herd and Chaudhari, 1974) provided no evidence of microcrystallites and the microdiffraction results suggest that translational symmetry, a prerequisite for crystallites, is not present even from diffraction areas as small as 10 Å in diameter (Geiss, 1976). The better agreement of the dense random-packed models over the microcrystallite models with experimental data and the lack of evidence by electron microscopy and microdiffraction for microcrystallites suggest that the short

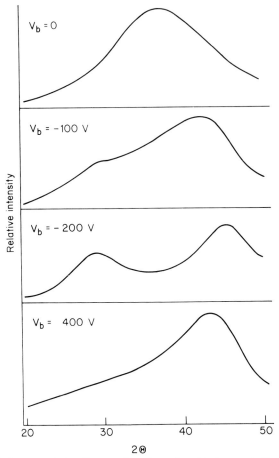

Fig. 11 Relative x-ray intensity as a function of scattering angle from a nominally $Gd_{28}Co_{72}$ film prepared as a function of bias voltage in the sputtering apparatus.

range order of amorphous Gd–Co alloys is best described in terms of non-crystalline models.

The preceding discussion was concerned with the average short-range atomic order in Gd–Co alloys. In amorphous solids there is no long-range atomic order. Beyond long-range atomic order is the question of microscopic homogeneity over distances of a few to hundreds of atom diameters. Put in another way, we are concerned with finding out if the amorphous films have electron density fluctuations over distances greater than a few atom diameters. Such fluctuations could be associated with second phases or voids. Both small angle scattering and transmission electron microscopy show that the evaporated and sputtered films contain electron density fluctuations (Cargill,

1977; Herd, 1977; Graczyk, 1977; Dirks and Leamy, 1977; Gill et al., 1977). In the evaporated films the experimental evidence appears to suggest columnar growth. The axis of the column follows the direction of the incident stream of atoms and is, for example, perpendicular to the plane of the film when the incident stream is normal to the plane of the substrate. The diameter of the column ranges from 30 to 100 Å depending upon the deposition conditions. The transmission electron diffraction pattern provides evidence of a void network on a scale of 100 Å. An electron diffraction pattern can be obtained when the electron beam is perpendicular and inclined to the plane of the film, and on tilting, the small angle scattering has anisotropic intensity distribution suggestive of anisotropic structure. Small angle scattering from bias-sputtered films shows similar anisotropic low angle scattering only from those films which are deposited at zero or low bias voltages. As the bias voltage is increased, the small angle scattering is decreased and shows a tendency towards isotropic scattering.

The large angle scattering, particularly in the vicinity of the first amorphous halo, shows changes with increasing bias voltage. The first amorphous halo splits into two haloes (Cargill, 1977) (see Fig. 11). Similar splitting of the first amorphous halo has been observed in films prepared by evaporation in an oxygen partial pressure and also in evaporated films annealed to high temperatures (Herd, 1977; Graczyk, 1977). The splitting of the first amorphous halo is suggestive of phase separation and has been associated with a mixture of gadolinium oxide and gadolinium–cobalt alloy (Wagner et al., 1976; Cargill, 1977). Attempts to correlate this splitting with magnetic anisotropy are discussed in a later section.

4 MAGNETIC MOMENT

Investigation of the properties of amorphous rare-earth transition metal alloys is only a few years old. It was largely triggered by the observation that sputtered films of Gd–Co and Gd–Fe can be prepared with a magnetic anisotropy perpendicular to the plane of the film. This anisotropy was sufficiently large that applications related to magnetic bubbles and beam addressable storage could be conceived (Chaudhari et al., 1973a,b). Since these early observations in Gd–Co, Gd–Fe, and Tb–Fe alloys (Chaudhari et al., 1973a,b; Orehotsky and Schroeder, 1972; Rhyne et al., 1974) a large number of binary and ternary alloys of rare earth–transition metal alloys have been examined. These include the rare earths Gd, Tb, Dy, Ho, Er, Tm, Yb, Lu, Nd, and transition metals such as Ni, Co, Fe, Mn, and Cr (Heiman and Lee, 1975b; McGuire et al., 1977b; Taylor et al., 1985; Schneider, 1975; Meyer et al., 1975; Minkiewicz et al., 1976; Sunago et al., 1976; Minamura

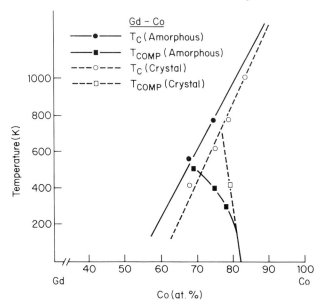

Fig. 12 The Curie and compensation temperatures of amorphous and crystalline Gd–Co alloys as a function of composition.

et al., 1976; Frait et al., 1976). In addition, a variety of non-magnetic ions have been added to the rare earth with the view of investigating the properties of just one magnetic ion in an amorphous milieu (Mizoguchi et al., 1977a). Similarly, Y has been added as a substitute for rare-earth atoms to transition metals such as Co and Fe (Suran et al., 1977; Kirshnan et al., 1977; Heiman and Lee, 1976). Ternary alloys where two rare earths with a transition metal or a single rare earth with two transition metals have also been investigated (Hasegawa and Taylor, 1975; Taylor, 1976a; McGuire et al., 1976). It is not our intent to review the entire field of amorphous rare earth–transition metal alloys but rather to confine ourselves to Gd-based alloys. By virtue of their relatively low coercivities Gd-based transition metal alloys have been the only class of materials which have been used in bubble device applications. It seems to us that this has been an unnecessary restriction on the use of rare earths other than Gd. Coercivities have generally been attributed to dispersion in the direction of single ion anisotropy associated with a non-S-state ion. If these dispersions arise from variations in the direction of the electrostatic field gradient in an amorphous solid then preparation or annealing in a magnetic field might be a useful way of aligning the single ion anisotropy, thereby increasing the uniaxial anisotropy and decreasing the coercivity.

Amorphous Gd–Co and Gd–Fe alloys are ferrimagnetic. The compen-

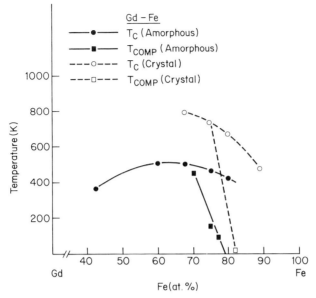

Fig. 13 The Curie and compensation temperature of amorphous and crystalline Gd–Fe alloys as a function of composition.

sation temperature is a function of composition, as is the moment per atom of the transition metal, and the Curie temperature (Heiman *et al.*, 1976). We show these variations with temperature and composition in Fig. 12 for the Gd–Co alloys. Corresponding values for the crystalline alloys are also shown in Fig. 12 for the moment per atom of the transition metal and the Curie temperature. The amorphous Gd–Fe alloys show a similar behavior but differ from the amorphous cobalt-based alloys in one marked aspect. The Curie and compensation temperatures of crystalline Gd–Fe are higher than in the amorphous Gd–Fe alloys. This is shown in Fig. 13.

Mean field theories have been used to describe the behavior of amorphous Gd–Co and Gd–Fe alloys (Hasegawa *et al.*, 1974a; Argyle *et al.*, 1974; Taylor and Gangulee, 1976). In these models it is assumed Gd and Co atoms are coupled antiparallel and form two subnetworks. The dominant exchange arises from the transition metal coupling and largely determines the Curie temperature. The moment per atom of Gd is assumed to be constant, independent of composition and structure. The moment of the transition metal is allowed to vary with composition and structure so as to satisfy experimental observations. The total magnetization is therefore written in the form

$$M_s = N\mu_B(x_1 g_1 \bar{S}_1 + x_2 g_2 \bar{S}_2), \tag{1}$$

where N is the number of atoms per unit volume; μ_B is the Bohr magneton; x_1 and x_2 are atomic fractions of Gd and Co; and g_1, S_1, and g_2, S_2 are the g-factors and spin values of Gd and Co, respectively. The value of $g_1\bar{S}_1$ is set equal to 7 and the product $g_2\bar{S}_2$ determined from the measured density and moment. The value of g_2 is obtained from ferromagnetic resonance yielding a value for the spin of the transition metal $\bar{S}_2$. This value for the spin is subsequently used to determine the temperature dependence of magnetization by fitting it to a Brillouin function, $Bs_i(x)$,

$$\bar{S}_i = S_i Bs_i(g_i S_i \mu_B H_i / kT), \quad (2)$$

where k is the Boltzmann constant, T is the temperature, and H_i is the effective field which includes the self field, the interaction field, and the applied field, H_a. We write

$$H_1 = 2J_{11}z_{11}\bar{S}_1/g_1\mu_B + 2J_{12}z_{12}\bar{S}_1/g_1\mu_B + H_a, \quad (3)$$

and

$$H_2 = 2J_{21}z_{21}\bar{S}_1/g_1\mu_B + 2J_{22}z_{22}\bar{S}_2/g_2\mu_B + H_a, \quad (4)$$

where J_{11}, J_{12}, and J_{22} are the exchange constants for Gd–Gd, Gd–Co, and Co–Co interactions, respectively. The numbers of nearest neighbors z_{11}, z_{12}, z_{21}, z_{22} can be obtained from the structural data as described in an earlier section. Traditionally, but not correctly, it is assumed that there are 12 nearest neighbors and therefore the values are $z_{11} = 12x_1$; $z_{12} = 12x_2$; $z_{21} = 12x_1$; $x_{22} = 12x_2$. The mean field fits to the data are usually quite good over the entire temperature range except perhaps close to the Curie temperature, where the experimental data lie at higher temperatures than calculated. A typical fit to the data is shown in Fig. 14. Once such a fit is made, the subnetwork magnetizations are determined and can be used to determine the exchange constant α, given by the equation (Hasegawa and Taylor, 1975; Gangulee and Taylor, 1977; Gangulee and Kobliska, 1985)

$$A = \sum_{ij} n_{ij} B_{ij} |J_{ij}| \bar{S}_i \bar{S}_j P_{ij} a_{ij}^{-1} \quad (5)$$

where a_{ij} is the interatomic distance between atom i and j, n_{ij} is the number of i–j nearest neighbor atom pairs per unit volume, and P_{ij} are given by $P_{11} = x_1^2$, $P_{12} = P_{21} = x_1 x_2$, and $P_{22} = x_2^2$.

The subnetwork magnetizations can also be used to determine the functional form of the subnetwork and temperature dependence of magnetic anisotropy. The temperature dependence of the subnetwork magnetizations can be directly obtained from experimental measurements, either by using the Kerr effects (Hasegawa et al., 1974a; Argyle et al., 1974; Taylor and Gangulee, 1976) or from ferromagnetic resonance studies (Chaudhari and Cronemeyer, 1976). In either case the absolute value of the subnetwork

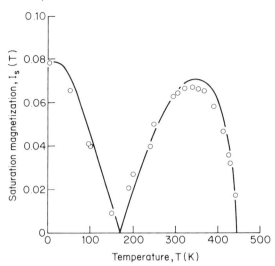

Fig. 14 Experimental and calculated saturation magnetization of an amorphous $Co_{0.67}Gd_{0.10}Mo_{0.14}Ar_{0.09}$ thin film. The solid line is calculated using mean field theory.

moment is not directly obtained. Moments from the Kerr rotation measurements are based on the assumption that the principal contribution to the Kerr rotation is from the cobalt subnetwork. This rotation is calibrated against a known cobalt moment. The Gd-subnetwork moment is obtained by subtracting the Co-subnetwork moment from the measured total moment. In ferromagnetic resonance studies the main resonance mode is measured with the applied field parallel and perpendicular to the plane of the film. Knowing the total moment of the sample, the effective gyromagnetic ratio g_e can be determined from the equation

$$g_e = (h/\beta)f\{[H_\parallel(1.25H_\parallel + H_\perp)]^{1/2} - H_\parallel/2\}, \qquad (6)$$

where h is Planck's constant, β is the Bohr magneton, f is the applied microwave frequency, and $H_\parallel$ and $H_\perp$ denote the resonance fields for a saturated state in the two configurations. The value of g_e can be related to the subnetwork moments through the Wangness expression

$$g_e = \frac{M_1 + M_2}{(M_1/g_1) + (M_2/g_2)}. \qquad (7)$$

As $M = M_1 + M_2$, we have two equations and two unknowns. If the value of g_1 and g_2 are known the temperature dependence of the subnetwork moment can be obtained by making the assumption that g_1 and g_2 are not functions of temperature. Knowledge of the subnetwork moments is desirable if we are to tune the magnetic properties and also to relate the magnetic anisotropy to subnetwork moments rather than the total moment which can lead to spurious interpretation.

One of the drawbacks of the binary alloy systems of amorphous Gd–Co has been the high magnetization for a given value of the fundamental length parameter, l. It is generally accepted, and certainly established in the case of T- and I-bar type devices, that a lower magnetization leads to a lower drive field (Kryder et al., 1974; George et al., 1974). The emphasis on bubble materials has been to lower magnetization without adversely affecting other magnetic properties. In one approach to lowering magnetization, "spin diluents" have been added to the Gd–Co alloys. Molybdenum addition is a good example of this class (Chaudhari et al., 1975). In another approach, the magnetization per unit volume is reduced by increasing the concentration of non-magnetic impurities or alloy diluents. For example, the Gd–Co–Mo–Ar system is both a spin diluent and alloy diluent system. Increasing argon content of the films results in a lowering of the magnetization without substantially changing the compensation and Curie temperatures (Cuomo and Gambino, 1977). An example of magnetization as a function of temperature, for various amounts of argon, is shown in Fig. 15. Experience with the Gd–Co–Mo–Ar system suggests that it is a useful submicrometer bubble material when viewed in context of the usual T- and I-bar devices. Devices that can tolerate larger moments and lower values of the quality factor Q can probably use the Gd–Co–Mo–Ar system at larger bubble

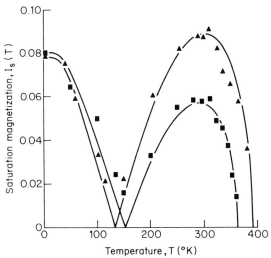

Fig. 15 Temperature dependence of the saturation magnetization of two Gd–Co–Mo–Ar films containing different amounts of Ar. The solid line is mean field fit.

diameters. Since a considerable amount of information is available in the published literature about the Gd–Co–Mo–Ar system, we shall use it to illustrate some of the important inter-relationships between bubble device requirements and materials properties. This will bring out both the limitations and the useful range of the amorphous bubble materials and, in particular, those of the Gd–Co–Mo–Ar system.

Satisfactory bubble materials must have an appropriate saturation magnetic moment $4\pi M_s$, the exchange constant A, perpendicular uniaxial magnetic anisotropy K_u, quality factor $Q \equiv K_u/2\pi M_s^2$, domain wall mobility μ, and coercivity H_c. In addition, commercial devices have to operate between room temperature and 80°C and therefore require that the temperature dependence of the magnetic properties should be as small as possible.

A commonly used interpretation in this laboratory of the temperature dependence of magnetic properties is to require that the bubble collapse field change, preferably monotonically, by no more than 20% over a 100°C increase above room temperature. In order to home into the useful range of bubble size where all of the requirements can be met, it is essential to have information on $4\pi M_s$, A, K_u, μ, H_c, and the temperature dependence of these properties as a function of composition. Once this is known, interpolation and extrapolation can be carried out. Mean field fitting has been used extensively to obtain the composition and temperature dependence of the saturation moment and the Co–Co and Co–Gd exchange energies. Over the composition range $x_{Co} = 0.61$–0.74, $x_{Gd} = 0.08$–0.17, $x_{Mo} = 0.11$–0.18, and $x_{Ar} = 0$–0.16, the cobalt spin S_{Co}, the exchange energy J_{Co-Co}, and J_{Co-Gd} could be related to the concentrations by the empirical expressions (Kobliska et al., 1977)

$$S_{Co} = 0.775 - 0.848(x_{Gd}/s_{Co})^{1.5} - 1.688(x_{Mo}/x_{Co}), \tag{8}$$

$$J_{Co-Co} = [2.329 - 2.737(x_{Gd}/s_{Co})] \times 10^{-14} \text{ erg}, \tag{9}$$

and

$$J_{Co-Gd} = (27.69 x_{Ar}^2 - 10.73 x_{Ar} - 2.374) \times 10^{-15} \text{ erg}. \tag{10}$$

The value of the exchange constant can be obtained from mean field theories and structural data. It has also been measured from spin wave observations using ferromagnetic resonance (Heiman and Lee, 1975a). The observed exchange constant is within a factor of two of the calculated value for that composition using the mean field theory. The magnetic anisotropy is a function of the deposition parameters and subnetwork moments. Experience suggests that over a restricted temperature range of 100°C above room temperature the magnetic anisotropy decreases approximately linearly with increasing temperature. The value of coercivity is acceptable and, as we shall

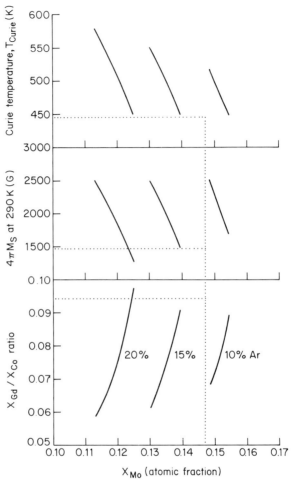

Fig. 16 Calculated values of Curie temperature, saturation magnetization, and X_{Gd}/X_{Co} as a function of Mo content in amorphous Gd–Co–Mo–Ar films.

discuss in detail in a later section, the values of mobility are more than adequate to meet device requirements. We concentrate therefore only on the static properties here.

As the Gd–Co alloys are ferrimagnetic, the overall objective is to determine compositions with peak values of the moment somewhat below room temperature. This ensures that the magnetization of the sample and the bias field decrease with increasing temperature. In order to demonstrate how the optimization procedure works, we select the peak of the magnetization curve to lie at 290°K. Using the empirical expressions we can calculate $4\pi M_s$, A,

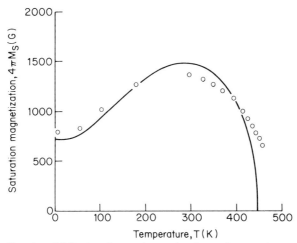

Fig. 17 Predicted (solid line) and experimental values of saturation magnetization as a function of temperature.

and the Curie temperature T_c as a function of composition. The results are shown in Fig. 16. The dotted line in this figure shows such an optimum composition if we fix the Curie temperature and the atomic fraction of Mo. The actual experimental and calculated curves for this composition are shown in Fig. 17. Typically the mean field prediction is within 10% of the experimental value in the interpolated compositions. Using a measured value of $Q = 2$ in these films and an exchange constant of 8.5×10^{-8} erg/cm at 295 K, the appropriate l value is 0.03 μm, which leads to a useful bubble diameter of approximately one-quarter micrometer.

5 ANNEALING AND ION IMPLANTATION

Annealing studies have been carried out with the view of determining the maximum processing temperature before irreversible changes in properties take place. Annealing and ion implantation studies have also been carried out both to study the origin of magnetic anisotropy and to tailor it.

Crystallization of the amorphous Gd–Co films leads to dramatic changes in magnetic properties of the films. In relatively thick films the crystallization temperature is in the vicinity of 500°C or higher. In thinner films, annealed in relatively poor vacuum, the crystallization temperature can be lower and in situ observations in an electron microscope suggest that it can be as low as 250°C. In thicker films or in flash-heated thin films, crystallization leads to Gd–Co compounds (Herd and Chaudhari, 1974; Herd, 1977; Graczyk, 1977; Hoffman and Höpfl, 1976). In thin films which are heated slowly, crystalline cobalt precipitates out. This is presumably due to selective oxidation of the rare earth.

Changes in magnetic properties can, however, occur at temperatures well below the crystallization temperature. It is these lower temperatures that are of interest from the standpoint of bubble applications. The temperature at which significant annealing changes occur is found to be a function of the composition of the rare earth–transition metal alloys. For example, in the Gd–Co–Mo alloys significant irreversible changes are observed above 250°C, whereas in the Gd–Co–Cu alloys irreversible changes in moment and anisotropy are already observed at temperatures around 100°C (Kobliska and Gangulee, 1975; Katayama et al., 1977). Magnetically, the biggest changes are observed to be in the compensation temperature and uniaxial anisotropy (Kobliska and Gangulee, 1975; Hasegawa et al., 1974b). A change in the compensation temperature can have a striking change in the magnetic moment at a given temperature unless the peak in the moment is at or close to this temperature. In the absence of crystallization we must conclude that the shift in the compensation temperature is not associated with long-range diffusion of the Gd–Co atoms. Changes in moment must either be associated with movement of impurity atoms or with local rearrangement of the atoms due to annealing out of defects such as local mass density fluctuations which in a crystal would correspond to vacancies. As the binary Gd–Co alloys anneal prior to crystallization the presence of Cu or Mo can only be concluded to affect the kinetics of annealing rather than lead to annealing. Also argon, which is incorporated during film deposition, is known to be released from the films at temperatures where massive crystallization is observed (M. Frisch, personal communication). We conclude that long range diffusion of argon is also not responsible for annealing of magnetic films. The most plausible mechanism appears to be a local rearrangement of atoms into energetically more favorable positions by removing structural inhomogeneities introduced during the evaporation or sputtering process. As the computer-simulation studies discussed in the structure section suggest atomic packing is anisotropic and contains "voids or cracks", we suggest that the magnetic annealing is a sensitive measure of such structural imperfection. As these anneal out there are slight changes in the average nearest neighbor environment. From a practical point of view, Gd–Co alloys which tend to anneal at temperatures lower than 100°C are undesirable both because of device processing complications and commercial device operation temperatures which can be close to, albeit lower than 100°C.

Ion implantation has been shown to change magnetic anisotropy (Gambino et al., 1974a; Venturini et al., 1976; Ali et al., 1976; Mizoguchi et al., 1977b). Such studies have been carried out to see if the implanted layer can be used,

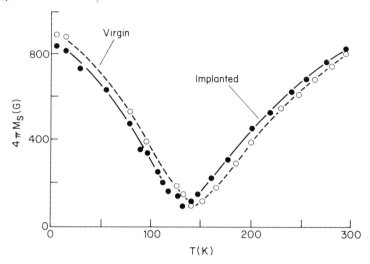

Fig. 18 Saturation magnetization versus temperature for an unimplanted (broken line) and implanted film (2×10^{15} 2-MeV Ar^+/cm^2).

as in garnets, as a hard bubble suppressing layer or as a drive layer in contiguous disc drives. We cannot at present assess the device implications of the implanted layer, since this work is still in progress. We shall therefore only discuss the effect of ion implantation and its significance in understanding the origin of magnetic anisotropy and its control.

Ion implantation experiments have been carried out in amorphous Gd–Co and Gd–Co–Mo alloys. An observation that was common to all of these experiments was a reduction in magnetic anisotropy with increasing implantation dose. As in annealing experiments, one also observes a shift in the compensation temperature and therefore a change in the saturation moment at room temperature. The magnitude of the change in moment at room temperature is a function of composition. We show in Fig. 18 the change in moment of a Gd–Co–Mo sample after an implantation dose of $2 \times 10^{15} Ar^+$ ions/cm². The energy of the incident ions was 2 MeV. The thickness of the Gd–Co–Mo film and the energy of the incident ions were selected to ensure that the probability of trapping any of the incident Ar^+ ions in the film was nearly zero.

The functional dependence of magnetic anisotropy on dose is shown in Fig. 19. Samples were irradiated at two different beam currents to ensure that the reduction in anisotropy was not associated with thermal heating of the sample during irradiation. The change in anisotropy is independent of ion beam current and is an exponential function of the integrated dose. This functional dependence can be explained by assuming that each ion generates

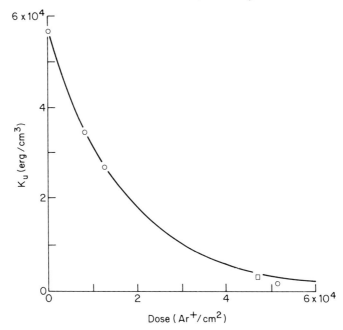

Fig. 19 Uniaxial anisotropy energy K_u, measured at room temperature as a function of implantation dose. Solid line is exponential fit to data. A square represents a datum for a film implanted with a higher ion current (5 μA) than others (0.5 μA).

a cascade of displaced atoms. Within each cascade volume the anisotropy is reduced by some fraction p. It is further assumed that the probability of implanting an already implanted region is proportional to the area of the implanted region. Hence the rate of change of anisotropy is governed by two factors (P. Chaudhari, personal communication; see also Hasegawa *et al.*, 1974b) the probability of implanting an unimplanted region and the fractional decay in anisotropy on implantation. Both of these considerations separately or together yield an exponential dependence on anisotropy of the form $K_u = K_{u0(\exp)} - S(1-p)N$, where K_u is the anisotropy after an integrated dose N, K_{uo} is the original anisotropy, and S is the average cross-section of damage per implanted ion. If it is assumed that $p = 0$—that is to say, in the volume of material where a cascade has occurred the displaced atoms contain no trace of the original anisotropy—we can determine S from the experimental data, and in the particular case of these Gd–Co–Mo samples $S = 58.4\,\text{Å}^2$ and the volume St, where t is thickness of film, equals $5.9 \times 10^5\,\text{Å}^3$ when $t = 1.1 \times 10^{-4}$ cm. If, on the other hand, we assume that $(1-p) = 10^{-2}$—that is, p is close to 1—we obtain for St a value of approximately $6 \times 10^7\,\text{Å}^3$. Radiation damage theories predict the number of displaced atoms for a 2-MeV argon ion in a solid with the average atomic number and weight of Gd–Co–Mo alloy to be approximately 4×10^4, which corresponds to a volume of approximately 6×10^6 Å^3. This number straddles the two numbers we had obtained using the two extreme values of p.

We now consider the implications of these numbers in elucidating the origin of magnetic anisotropy. The damage track in a single crystal solid consists of clumps of displaced atoms. The number of such clusters cascades away from the initial knock-on. If we assume that there are 10 clusters per argon ion then the volume per cluster is between 6×10^5 and 6×10^7 Å^3. Assuming further a spherical cluster, the radius of the cluster is between 24 and 110 Å. On the basis of these results we can place an upper limit to the radius of structural and compositional fluctuation. It must be less than 250 Å, which is the largest radius possible assuming the entire damage is localized to one cluster and not distributed over 10 and the value of $(1 - p) = 10^{-2}$. We believe a more reasonable radius over which compositional and structural inhomogeneities might exist is closer to 50 Å rather than 250 Å. It is, of course, important to recognize that the ion implantation results are not inconsistent with magnetic anisotropy associated with short ranged atomic ordering. We shall return to this discussion again in the next section.

6 MAGNETIC ANISOTROPY

Magnetic anisotropy in amorphous materials can be sufficiently large that stable magnetic bubbles can exist. The anisotropy energy is larger than the demagnetizing energy. In this section we shall be discussing the origin of this anisotropy in amorphous materials. We briefly recapitulate the relationship of anisotropy to processing parameters and to atomic structure before proceeding to elaborate on the models proposed to explain the origin of anisotropy. Magnetic anisotropy is observed to increase with increasing bias fields in rf sputtering of Gd–Co–Mo films. Beyond a certain bias field, the anisotropy begins to decrease again. Changing bias also changes the composition of the film, both by changing the metal atom ratios and by increasing the argon (or other inert gas) content. In fact, the argon content shows a similar dependence on bias as the magnetic anisotropy (Cuomo and Gambino, 1977; Burilla et al., 1977) and is the basis of the sputtering model of anisotropy (Gambino and Cuomo, 1978). On a structural level, increasing bias tends to decrease the anisotropic low angle scattering of electrons and x-rays and, at higher bias voltages, tends to reintroduce low angle scattering, which is, however, isotropic. Low angle scattering is generally associated with structural inhomogeneities (to be precise with electron density inhomogeneities) of the order of 100 Å or smaller. In the large angle scattering data, which is a measure of atomic scale structure, increasing bias leads to the appearance of a split first peak. This has been associated with phase separation in the amorphous films. However, the shape of the two phases, as the small angle data suggest, is isotropic.

In evaporated films introduction of oxygen or nitrogen in the vacuum system leads to the development of a perpendicular component of anisotropy in the Gd–Co system (Brunsch and Schneider, 1977; Schneider and Brunsch, 1977). Incorporation of these gases also leads to a shift in the ratio of Gd to Co for the room-temperature compensation point. This is generally interpreted to result from the preferential reaction of Gd with the atoms of the gas. Increasing the gas pressure increases the perpendicular anisotropy until the material becomes superparamagnetic at partial pressures of oxygen in the range of 10^{-7} torr. Electron diffraction and transmission micrographs show that with increasing oxygen content the first amorphous halo appears to sharpen and split into two haloes, as in the case of large-bias voltage sputtering. There is anisotropic small angle scattering in as-deposited films associated with a void network in the film. This small angle scattering is independent of the oxygen content. Annealing of evaporated thin films prepared with a low background of oxygen partial pressures and therefore low perpendicular anisotropy shows an increase in perpendicular anisotropy prior to crystallization. The evolution of perpendicular anisotropy is accompanied by first the disappearance and at higher temperatures the reappearance of anisotropic low angle scattering and by splitting of the first amorphous halo. The disappearance of the low angle scattering is associated with the annealing out of voids and its reappearance with phase separation (Cargill, 1977; Herd, 1977; Graczyk, 1977).

Films of Gd_x–Co_{1-x} ($x \gtrsim 0.1$) evaporated in low oxygen pressure prefer to have an easy axis in the plane of the film (Heiman *et al.*, 1974; Onton and Lee, 1976; Taylor, 1976b). However, films of Gd–Fe and other rare earth–transition metal alloys generally show a perpendicular component of magnetic anisotropy (Heiman and Lee, 1974). Ternary Gd–Co–Fe films show perpendicular anisotropy which decreases with increasing Gd and Co content (Suran *et al.*, 1977; Kirshnan *et al.*, 1977; Heiman and Lee, 1976; Taylor, 1976a). Structural details of these films are not available.

The magnetostriction constants of Gd–Co alloys are generally too low to explain the magnitude of the observed anisotropy in thin films even if we assume that biaxial stresses of the order of 10^{10} dyn/cm^2 are present. Furthermore, the Gd–Co films have been examined when they are not constrained by the substrate. The magnitude of anisotropy is not significantly changed. The usual stress-induced anisotropy is probably not significant in these films. The Gd–Fe alloys have larger magnetostriction constants and

Amorphous Gd–Co Alloys

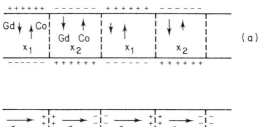

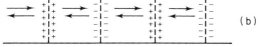

Fig. 20 Schematic distribution of poles when the composition fluctuates spatially about the compensation composition: (a) magnetization is perpendicular to plane of film and (b) magnetization is in-plane.

the Gd–Co–O alloys may have unusual microstructure. It is possible that in some of these alloys stress-induced anisotropy may contribute significantly (Brunsch and Schneider, 1977; Schneider and Brunsch, 1977; Zwingman *et al.*, 1976).

Attempts to explain the origin of uniaxial perpendicular anisotropy have taken two broad approaches. Structural anisotropy is emphasized in one of them on a scale of tens of atom diameters (shape anisotropy) (Cargill, 1977; Herd, 1977; Graczyk, 1977; Zwingman *et al.*, 1976; Cargill and Mizoguchi, 1977) and in the other on a scale of atomic dimensions (pair or short-range atomic ordering) (Chaudhari *et al.*, 1973a; Cargill and Mizoguchi, 1977; Esho and Fujiwara, 1976; Gambino *et al.*, 1974b). We consider first how structural fluctuations can give rise to magnetic anisotropy and examine this model in light of the available experimental data. Shape anisotropy arises from the anisotropic demagnetizing energy associated with a shaped magnetic region. Voids or composition fluctuations can give rise to such shapes. For example, if a thin film is composed of rod-shaped materials with the rod diameter less than the film thickness, a perpendicular component of magnetic anisotropy is observed. Similarly, if there are composition fluctuations in the plane of the film a perpendicular component of magnetic anisotropy can develop. The origin of this is shown schematically in Fig. 20. Figure 20a shows the case where the net magnetization is pointing perpendicular to the plane of the film and the demagnetizing energy is associated with the free surface. In Fig. 20b the magnetization is in-plane and now the free poles are generated along a surface where the magnetization changes spatially. If we assume that the fluctuations in magnetization can be represented spatially in the x–y plane (plane of the film) by the expression (J. Slonczewski, personal communication)

$$M = M_0 + M_1 \cos k_x x \cos k_y y, \qquad (11)$$

where k_x and k_y are wave vectors describing the wavelength of the magnetization fluctuation, then the anisotropy is given by the expression

$$K_u = \frac{\pi}{4} M_1^2 \qquad (12)$$

or more generally by the expression (J. Slonczewski, personal communication)

$$K_u = \pi M_{rms}^2. \qquad (13)$$

Denoting the magnetization in adjacent areas by M' and M'' we can write $Q(\equiv K_u/2\pi M_s^2)$ as

$$Q = (\bar{M}' - \bar{M}'')^2 / 2(\bar{M}' + \bar{M}'')^2 \qquad (14)$$

This expression tells us that Q can be greater than 1 if $\bar{M}'$ and $\bar{M}''$ are opposite in sign. In a ferrimagnetic system this is possible if the neighboring areas are exchange-coupled and also fluctuate in composition so as to straddle the compensation point. Contrary to ferromagnetic systems a ferrimagnetic material can have $Q > 1$ where the anisotropy is associated with a shape effect (Cargill and Mizoguchi, 1977).

We can obtain some idea of the magnitude of the composition fluctuations required to produce observable anisotropies by fitting the observed temperature dependence of anisotropy with approximate mean field fits to the data.

Using mean field theory we can write the approximate magnetization of the two regions as [see Eq. (1)]

$$M' = c[x'_1 g'_1 \bar{S}'_1 + x'_2 g'_2 \bar{S}'_2] \qquad (15)$$

and

$$M'' = c[x''_1 g''_1 \bar{S}''_1 + x''_2 g''_2 \bar{S}''_2] \qquad (16)$$

whose difference is $(M' - M'') = c[(x'_1 g'_1 \bar{S}'_1 - x''_1 g''_1 \bar{S}''_1) - (x'_2 g'_2 \bar{S}'_2 - x''_2 g''_2 \bar{S}''_2)]$. Expressing $\langle x_1 \rangle = (x'_1 + x''_1)/2$ and $x'_1 - x''_1 = 2(\Delta x)$ where (Δx) is a small quantity, we can rewrite the expression for $(M' - M'')$ in the form

$$(M' - M'') = c(1\Delta x)[\langle g_1 \rangle \langle \bar{S}_1 \rangle - \langle g_2 \rangle \langle \bar{S}_2 \rangle] \qquad (17)$$

where we have kept only the dominant terms in the expansion of the quantity in brackets of Eqs. (15) and (16). Equation (17) can be rewritten in the convenient form

$$(M' - M'') = (2\Delta x) \left[\frac{\langle M_1 \rangle}{\langle x_1 \rangle} + \frac{\langle M_2 \rangle}{\langle x_2 \rangle} \right] \qquad (18)$$

where $\langle \cdot \rangle$ indicate averages and relate directly to the mean field fits of subnetwork moments to the magnetization data and compositional analysis.

Amorphous Gd–Co Alloys

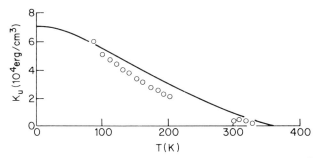

Fig. 21 Temperature dependence of uniaxial anisotropy energy as a function of temperature for an amorphous Gd–Co film. Solid line is calculated.

Equation (18) in conjunction with Eq. (13) has been used to fit the measured anisotropy data as a function of temperature. The subnetwork moments were obtained from the mean field fits to the net magnetization. Agreement between the experimental and calculated fits is usually quite good (see Fig. 21) and the value of Δx is typically of the order of 0.05. The composition fluctuation model therefore requires a fluctuation in composition of the Gd–Co ratio by $\pm 5\%$ to explain the observed anisotropy. It requires the composition fluctuations to be in the plane of the film.

Structural data in support of this kind of shape-induced anisotropy is available only from the evaporated thin films examined in transmission electron microscopy. We recall that films prepared in an oxygen environment and also the annealed thin films show evidence of anisotropic small angle scattering and splitting of the first amorphous halo. In contrast to the evaporated films, the sputtered films show splitting of the first amorphous halo but no evidence of anisotropic small angle scattering. One reason may be that a few percent composition change produces only a small change in the low angle intensity profile which may not have been detected.

Two additional experiments have been tried to determine the presence of regions where the net moment is antiparallel. The first of these has been high-field Hall effect measurements and the second is susceptibility measurements near the compensation point of the material. The sign and magnitude of the extraordinary Hall effect has been suggested to be a measure of the difference in scattering by the Gd and Co atoms. At fields where the sample is saturated, the spins of the cobalt subnetwork point in one direction and those of gadolinium point in the opposite. The net magnetization points in the direction of the external field. As the external field is increased, the Zeeman energy eventually overcomes the exchange energy holding the subnetwork. When this happens we expect the Hall signal to change. The external field at which this change is observed is a measure of the size of the two regions where the local net magnetizations are initially opposed in direction. A rough estimate of the field required to flip over two such regions can be calculated by assuming that the exchange energy is localized to the interface region of width equal to a. The exchange energy is $2\pi RwA/a$, where R is the radius of the first region, w is the thickness of the film, and A is the exchange constant. The Zeeman term is $-2M'H\pi R^2 w$ where M' is the magnetization of the second region which flips in field H. Equating the two terms and minimizing with respect to R gives

$$H_c \lesssim \frac{A}{2M'aR}. \tag{19}$$

Typical values of A are in the neighborhood of 10^{-7} erg/cm, $a \approx 3 \times 10^{-8}$ cm, and M' is probably around 100 G, yielding a value of H_c in the neighborhood of 10^4–10^5 Oe for a radius R lying between 20 and 100 Å. Attempts to measure this flip-over in fields of up to 10^5 Oe have been unsuccessful (G. S. Cargill, III, P. Chaudhari, and R. J. Gambino, unpublished observations). The Hall signal is essentially unchanged once saturation of the sample is achieved. These measurements suggest that the wavelength of the fluctuation, if present, must be less than 30 Å. It would be useful to measure the Hall coefficient at even larger fields. An alternate approach to looking for fluctuation where the net magnetization oscillates in sign is to measure susceptibility at the compensation point of the sample. This susceptibility is given by (Mizoguchi et al., 1977c)

$$\chi = M/H = M'^2 R^2 / \pi^2 A. \tag{20}$$

Substituting for M', R, and A values of 20 G, 50×10^{-8} cm, and 10^{-7} erg/cm, we find $\chi = 2.5 \times 10^{-3}$. Rather than use these values, which have some uncertainty in them, we can use the high-field measurements to obtain an upper limit to susceptibility. We noted that the Hall signal did not show any spin flipping up to fields of 10^5 Oe. We estimate that the ratio $M'R/A$ must therefore be less than 1.7×10^2. Using an A of 10^{-7} erg/cm we find χ to be less than 3×10^{-4}. The measured value of χ is of the order of 2.5×10^{-4}. Both the high-field and susceptibility data suggest that shape-induced anisotropy, if present, is associated with composition fluctuations which span distances of the order of 30 Å or smaller. As we have remarked before, anisotropic composition fluctuations of this kind have been observed in certain evaporated films but not in bias-sputtered amorphous films. We must therefore conclude that there is no experimental evidence in favor of shape-induced anisotropy in sputtered films where the linear dimension of the shape is greater than 30 Å. Experimental evidence does not, at present, provide us with unequivocal information about shape effect contributions arising from shapes of dimensions 30 Å or smaller in sputtered films.

We now consider the second possible origin of magnetic anisotropy. This is associated with pair ordering or more generally dipolar-induced anisotropy. The classical dipolar interaction between two magnetic atoms i and j which are locally in parallel ($+$) or antiparallel ($-$) arrangement is given by (Cargill and Mizoguchi, 1977)

$$U_{ij} = \pm \frac{m_i m_j}{r_{ij}^3}[1 - 3(\alpha_1\beta_1 + \alpha_2\beta_2 + \alpha_3\beta_3)] \tag{21}$$

where α and β are direction cosines of the spin i (or j) and the vector r_{ij}, respectively, with respect to a laboratory frame of reference.

In an isotropic amorphous solid, the direction cosine term averaged over all orientations yields a value of $\frac{1}{3}$ so that the average dipolar interaction is zero. If we assume that the dipole distribution is anisotropic, then there is a resultant dipolar-induced anisotropy. To a first approximation this can be written as (Cargill and Mizoguchi, 1977)

$$K_u = C\left[\frac{\langle M_1 \rangle}{\langle x_1 \rangle} + \frac{\langle M_2 \rangle}{\langle x_2 \rangle}\right] \tag{22}$$

and

$$C \approx 0.42 z p x_1 x_2, \tag{23}$$

where p is a measure of the degree of anisotropy distribution of dipole pairs. As Mizoguchi recognized for a dipolar interaction and Cargill in the case of shape-induced anisotropy, a ferrimagnetic alignment can lead to $Q (\equiv K_u/2\pi M_s^2) > 1$ since $M_s^2 = (|M_1| - |M_2|)^2$ and K_u is proportional to $[|(M_1/x_1)| + |(M_2/x_2)|]^2$. The classical dipolar interaction can be generalized to include anisotropic exchange interaction, in which case K_u is written as (Chaudhari and Cronemeyer, 1976)

$$K_u = \sum_{ij} C_{ij} M_i M_j, \tag{24}$$

where the C_{ij} are pseudo-dipolar coefficients. The pseudo-dipolar equation has been shown to fit a variety of amorphous Gd–Co alloys. We show an example of the pseudo-dipolar fit to the anisotropy measured as a function of temperature in Fig. 22. The subnetwork moments were deduced from the measured or effective gyromagnetic ratio obtained from ferromagnetic resonance data. Other than the fitting of the data, there is no direct structural data in support of the dipolar model. A part of the reason why no anisotropy in structure is observed by diffraction techniques may lie with the very small value of p. Diffraction techniques are currently not sensitive enough to pick up this anisotropy in pair distribution, if it is present.

It is apparent from our discussion that the origin of anisotropy is still not precisely known. In sputtered films either a dipolar model or a shape-induced

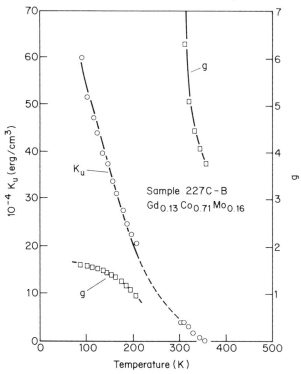

Fig. 22 Temperature dependence of uniaxial anisotropy in an amorphous Gd–Co alloy. Solid line is pseudo-dipolar fit.

effect, where the shape is less than 30 Å in diameter, could explain the magnetic anisotropy. There is no direct structural evidence in support of one model or the other. In the evaporated films, structural evidence for a shape effect is present. However, more magnetization measurements are needed to establish that this structural anisotropy is indeed the major source of magnetic anisotropy.

7 DOMAIN WALL DYNAMICS OF AMORPHOUS FILMS

Most of the data on the dynamic properties of amorphous films have come from the Gd–Co–Mo alloy system. Some data are also available for Gd–Co, Gd–Co–Au, and Gd–Co–Cu samples. Dynamic measurements on the amorphous films have been made by a number of techniques, and the data are quite consistent. At present, however, there are some features of the data which are not well understood. Kryder and Hu (1974) first reported

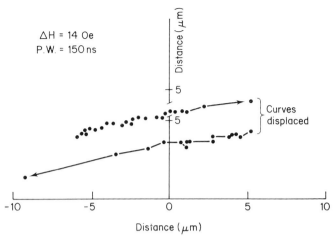

Fig. 23 Bubble coordinates after each of a series of pulses. Vertical axis is parallel to and centered between conductors. Bubble diameter is 1 μm. (From Potter et al., 1976.)

measurements of the dynamic properties on amorphous films of Gd–Co, Gd–Co–Au, and Gd–Co–Cu by the bubble translation technique first described by Vella-Coleiro and Tabor (1972). In a Gd–Co film these measurements revealed velocities as high as 100 m/sec under conditions of very high drive, and in a Gd–Co–Au film velocities in excess of 50 m/sec were found. In performing these measurements, Kryder and Hu did not use bias field compensation; hence the bubble was caused to expand and possibly stripout during the translation. As Josephs and Stein (1975) and Potter et al. (1976) showed, this can lead to anomalous values for mobility and dynamic coercivity. An example of the data of Potter et al. are shown in Fig. 23. The data indicate that consistently small displacements are measured if the bubble is propagated towards the center line between the strip conductors so that a positive bias field pulse accompanies the gradient pulse. However, if the bubble is propagated away from the center line, so that a negative bias field pulse accompanies the gradient pulse, displacements up to one order of magnitude larger are found. Nevertheless, in both the Kryder and Hu and Potter et al. measurements, velocities larger than 50 m/sec were found when the bubble propagated away from the center line, indicating that the saturation velocities in 2-μm amorphous films are indeed very high. Having discovered the large variation in velocity as a function of position with respect to the center line, Potter et al. were careful to determine the mobility and dynamic coercivity only from measurements of the velocity as the bubble approached the center line between the drive conductors. Thus their data on mobility are expected, if anything, to be conservatively low.

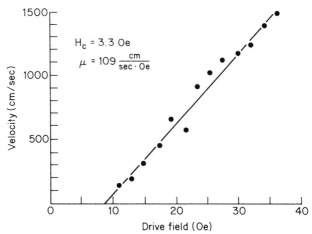

Fig. 24 Velocity versus drive field ΔH for a representative unannealed sample. (From Potter et al., 1976.)

An example of their data is shown in Fig. 24 and a summary of their findings is included in Table 1 along with data from a number of other authors. As seen from Table 1, they obtained mobilities ranging from 0.92 to 3.25 m/sec Oe and dynamic coercivities ($8H_c/\pi$) of 5.3 to 14.1 Oe for nominally 1 μm stripwidth Gd–Co–Cu materials with $4\pi M_s \approx 1500$ G.

Both Kryder and Hu (1974) and Potter et al. (1976) found evidence of Bloch lines in Gd–Co, Gd–Co–Au, and Gd–Co–Cu samples. Kryder and Hu observed hard bubbles with anomalously high collapse fields, dumbbell domains which would rotate due to gyrotropic forces (Slonczewski et al., 1973) in a pulsed field, and bubbles which were deflected from the gradient field direction. Potter et al. showed that the deflection angles in Gd–Co–Cu were quantized as Slonczewski et al. (1973) had shown in garnets.

The dynamic properties of the Gd–Co–Mo system have been determined using bubble translation, bubble and stripe expansion, and a number of device measurements at high frequency by Kryder et al. (1977), Hafner and Humphrey (1977), and R. I. Potter (personal communication). A summary of their measurements is also included in Table 1. All these measurements indicate approximately consistent values for mobility and velocity in Gd–Co–Mo films of similar stripwidth.

Bubble translation in a pulsed gradient field was used by both Kryder et al. (1977) and R. I. Potter (personal communication) to determine the mobility of Gd–Co–Mo films. Both were careful to avoid stripout effects. Figure 25 shows a plot of bubble translation velocity versus gradient drive field taken by Kryder et al. (1977) on a 2-μm stripwidth sample. The data yield a wall mobility of 2 m/sec·Oe and a dynamic coercivity ($8H_c/\pi$) of

Table 1 Dynamic properties of amorphous film samples.

Sample	Reference	Composition	Thickness (μm)	Stripwidth (μm)	$4\pi M_s$ (G)	H_k (Oe)	H_c(Oe) Static	H_c(Oe) Dynamic	μ_w (m/sec·Oe)	V_{SAT} (m/sec)
144	Kryder	Gd–Co	1.7	1.8	1100			6.3*	20*	>110
39c	Kryder	Gd–Co–Au	1.7	2.3	1000			1.0*	14.3*	~50
	Potter	Gd–Co–Cu	2.29	1.5	1460			5.0	2	
	Potter	After annealing	2.29	1.02	1750			2.1	3	
	Potter	Gd–Co–Cu	1.24	1.1	1060			Unmeasurable	Unmeasurable	
	Potter	After annealing	1.24	0.9	1390			3.8	3.25	
	Potter	Gd–Co–Cu	1.24	1.1	930			Unmeasurable	Unmeasurable	
	Potter	After annealing	1.24	0.9	1600			2.0	0.92	
	Potter	Gd–Co–Cu	2.29	1.6				3.8	2.05	
	Potter	After annealing	2.29					5.8	1.7	
BBL15-1	Kryder	Gd–Co–Mo	2.04	2.19	505	1750			2.5	~100
BBL222-1	Kryder	Gd–Co–Mo	1.85	2.27	265	1725			1.2	
BJL 43-3	Kryder	Gd–Co–Mo	2.00	1.9	532	2100			2	
9UA	Kryder	Gd–Co–Mo	1.38	2.0	447	1200		1.7	2	
61-28-15	Potter	Gd–Co–Mo	1.21	1.28	518	2276	2.3	4.5	0.74	
61-29-5	Potter	Gd–Co–Mo	1.29	1.24	556	2403	2.9	4.7	0.84	
61-27-11	Potter	Gd–Co–Mo	1.21	1.05	551	1975	2.5	4.7	0.88	
56-26-1	Hafner	Gd–Co–Mo	2.0	4.5		2400	6.4		1.95	>400
56-30-3	Hafner	Gd–Co–Mo	2.0	3.2		2200	10	10		~120

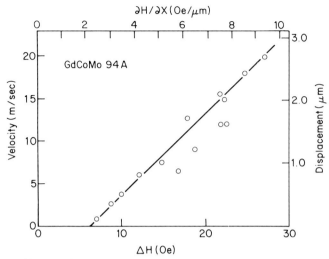

Fig. 25 Bubble displacement versus gradient field amplitude for 134-nsec field pulse. Also shown is calculated velocity drive field across bubble diameter. (From Kryder *et al.*, 1977.)

6.5 Oe. An example of R. I. Potter's data (personal communication) for a 1.28-μm stripwidth sample is shown in Fig. 26. Here $\mu_\omega \approx 0.75$ m/sec·Oe and $8H_c/\pi \approx 18$ Oe. The data of Kryder *et al.* (1977) and R. I. Potter (personal communication) indicate typical values of mobility of about 2 m/sec·Oe for 2-μm stripwidth samples and of about 0.75 m/sec·Oe for 1-μm stripwidth samples. The large dynamic coercivities found by Potter for a large number of 1-μm samples are not understood, however. The quasi-static coercivities

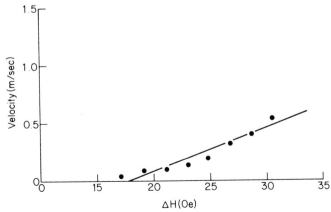

Fig. 26 Calculated velocity versus drive field across the bubble diameter. (From R. I. Potter, personal communication.)

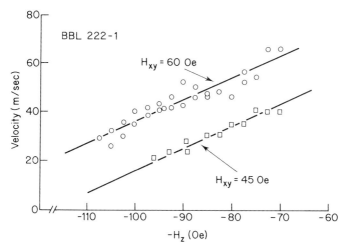

Fig. 27 Chevron expansion velocity versus H_z for device BBL 222-1. (From Kryder et al., 1977.)

determined from hysteresis loops for these 1-μm stripwidth films are typically a factor of two lower than those in 2-μm stripwidth films and a factor of three lower than the dynamic coercivities found by Potter.

An interesting finding from the bubble translation measurements on Gd–Co–Mo films is that the films exhibit no evidence of hard bubbles. No bubbles were observed to propagate at an angle to the gradient field, and no bubbles were found to have anomalously high collapse fields. This is in direct contrast to the garnets and the other amorphous film alloy systems already mentioned. Measurements on films in which the oxidized surface layer was removed and in situ coated with SiO_2 also exhibited this feature. Although it is still possible that a very thin (<200 Å) in-plane layer exists at the substrate–film or film–SiO_2 overcoat interface, it is not expected that such a thin layer would be so effective in preventing hard bubbles. The absence of hard bubbles in Gd–Co–Mo films is a yet unexplained but fortunate finding.

Measurements of bubble expansion velocity on chevron expander structures with a high speed magneto-optic camera system by Kryder et al. (1977) yielded mobilities similar to those measured by bubble translation at low drive, but unusually high velocities were found under high drive conditions. Figures 27 and 28 show the Kryder et al. (1977) data for chevron expansion velocity as a function of decreasing bias field. The data in Fig. 27 indicate a mobility of 1.2 m/sec·Oe, independent of any in-plane rotating field. The data in Fig. 28 for another device indicate a mobility of 2.5 m/sec·Oe for low in-plane fields (<50 Oe), but unusually high velocities and non-constant

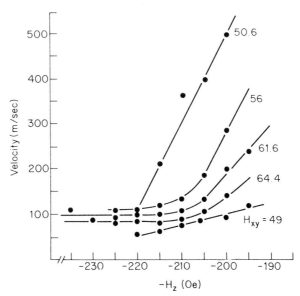

Fig. 28 Chevron expansion velocity versus H_z for device BBL 15-1. (From Kryder et al., 1977.)

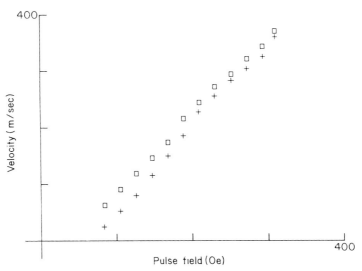

Fig. 29 Average velocity during the first 100 nsec of expansion of a magnetic bubble (□) and stripe (+) wall as a function of the applied pulse field for sample 56-21-1. (From Hafner and Humphrey, 1977.)

mobilities were found at high in-plane fields (>50 Oe). The fact that such behavior was not found in the device of Fig. 27 is attributed to the fact that the spacer between the film and the NiFe chevron elements was much thicker (3500 Å) in that device than the spacer (1500 Å) in the device used to obtain the results in Fig. 28. The thicker spacer led to reduced drive from the permalloy and the region of high velocity was not reached.

The data of Hafner and Humphrey (1977) on bubble stripe expansion as a function of drive field, also measured with a high-speed magneto-optic camera system, provide additional confirmation of the mobilities and velocities measured by Kryder *et al.* (1977) and R. I. Potter (personal communication). Two plots of these data are shown in Figs. 29 and 30. The data in Fig. 29 were taken on a film with 4.5 μm stripwidth and shows a constant slope of about 1.3 m/sec·Oe up to velocities of 400 m/sec. Hafner and Humphrey (1977) point out that the field driving the bubble wall during their experiment depends not only upon the applied pulse field but also the transient bubble size. By taking both these factors into account, they determined a mobility of 1.95 m/sec·Oe for sample 56-26-1 of Fig. 29. The data in Fig. 30 were also taken by Hafner and Humphrey (1977) on a film with 3.2 μm stripwidth and shows a low-field slope of 1.5 m/sec·Oe and a saturation at about 135 m/sec, which is followed by a subsequent slow rise in velocity. These data are in many ways similar to that of Fig. 28. Figure 28 shows a low-field mobility of 2.5 m/sec·Oe with 49 Oe and lower rotating

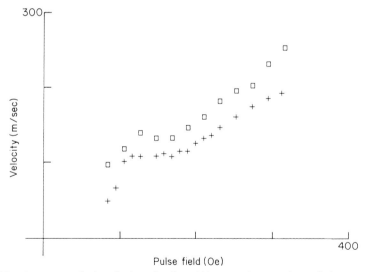

Fig. 30 Average velocity during the first 100 nsec of expansion of the magnetic bubble (□) and stripe (+) wall as a function of the applied pulse field for sample 56-30-3. (From Hafner and Humphrey, 1977.)

drive fields. At about 100 m/sec there is a velocity saturation but when the bias is decreased sufficiently (causing high drive toward stripout), gradually increasing velocities are observed. The only feature not contained in Fig. 29 is the velocity peak at 500 m/sec which occurs with the 50.6 Oe rotating drive field in Fig. 28. Since this peak is extremely narrow, and may furthermore depend in some critical way on the combination of in-plane rotating field and the bias field, it is certainly possible that it exists also in Hafner and Humphrey's (1977) film but was not found with their techniques.

In summary, the dynamic properties of the amorphous films under low drive conditions appear therefore to be relatively well characterized. By a large variety of techniques it has been established that 2-μm stripwidth Gd–Co–Mo films have mobilities of about 2 m/sec·Oe, whereas 1-μm stripwidth films have mobilities of about 0.75 m/sec·Oe. These values of mobility are not large but are certainly sufficient for good device operation up to 1 MHz, as detailed later in the device section where device performance at high frequencies is discussed at length.

In spite of the well established values of mobility, no fundamental explanation of damping mechanism in amorphous films has been proposed which provides reasonable fit to even the low-drive mobility data. Kryder and Hu (1974) considered the case of eddy current damping, and showed that for an infinitely long straight domain wall the eddy current limited mobility is given by

$$\mu_w = \frac{c^2 \rho}{32\pi \log 2hM} \tag{25}$$

where h is the film thickness, M is the saturation magnetization, ρ is the resistivity, and c is the speed of light. For the Gd–Co–Mo films under consideration here, $M = 40$ emu, $h = 2 \times 10^{-4}$ cm, and $\rho = 2 \times 10^{-4}\,\Omega\cdot$cm, yielding $\mu_w \approx 8 \times 10^3$ m/sec·Oe. This value is several orders of magnitude higher than actually found in the amorphous films and it can be concluded that eddy currents are not the dominant damping mechanism. Some other mechanism is required to explain the measured mobilities.

The non-linear dynamic behavior of the Gd–Co–Mo amorphous films under high-drive conditions is also not yet understood. It seems probable that the saturation and subsequent increase may be related to instabilities in the dynamic processes such as those discussed by B. R. Walker (unpublished observations, cited in Dillon, 1973), Slonczewski (1973), and Hagedorn (1974). Indeed, the data available thus far neither prove nor disprove correspondence with the various models which have been put forth for explaining the dynamic effects in bubble materials. More data on the gyromagnetic ratio, the damping parameter, and the dynamic behavior of the amorphous films is required before these features can be understood.

8 AMORPHOUS FILM BUBBLE DEVICES

Amorphous films have been considered as candidates for three types of bubble devices: (1) T-bar (Bobeck and Schovill, 1971) or half-disk (Bonyhard and Smith, 1976) field accessed devices using permalloy structures and a rotating field for driving the bubbles, (2) contiguous disk devices (Lin et al., 1977) using ion-implanted structures and a rotating field for driving the bubbles, and (3) bubble lattice file devices (Voegeli et al., 1975) using current-carrying conductors for driving the bubbles. Thus far amorphous films have been reported to be used only for T-bar devices. Although, as discussed earlier, ion implantation has been found to produce measurable effects on amorphous films, ion-implanted contiguous disk devices have not been fabricated on them. Amorphous films have not found use in the bubble lattice file for two reasons. First, the coercivity of the films is high, and prohibitively high currents would be required in the current-carrying conductors used to translate the lattice and to access a column. Secondly, in the Gd–Co–Mo system all bubbles have been observed to propagate approximately parallel to the direction of the field gradient; thus the bubble coding scheme presently popular for the lattice devices would not function. The remainder of the section on devices will therefore be directed to the more conventional devices using permalloy structures to drive the bubbles.

The majority of work on amorphous film devices reported in the literature has been on designs requiring no critical mask alignments. Bobeck et al. (1973) first reported single-mask-level bubble devices made on garnets. The majority of what they described referred to NiFe-only designs in which the NiFe was used as propagation elements, conductors, and detectors. In large major–minor loop chips with lengthy current lines it is impractical to use NiFe as conductor lines for block transfer switches because the voltage drop and power dissipation become prohibitively large. To overcome this problem, Kryder et al. (1976) used a layered structure of NiFe and Au in place of the NiFe only, as shown in Fig. 31a. To prevent the Au from shorting the magneto-resistive detector signal, a second masking step with extremely wide ($\pm 25\,\mu$m) alignment tolerance was used to prevent deposition of Au in the sensor area. Although this structure reduces the resistance in the conductor lines, some of this advantage is lost because the NiFe acts as a shield for the magnetic flux generated by the current in the conductor. As a result, higher currents are required in this NiFe-first, Au-second structure. Figure 31b shows an Au-first, NiFe-second structure which is more favourable with respect to the current requirements. In this case the NiFe acts as a keeper, intensifying the fields in the bubble material. Measurements on identical switch designs made by the Au-first and Au-second processes indicate

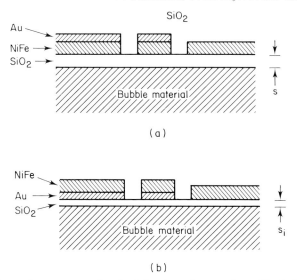

Fig. 31 Alternative structures for single level mask amorphous film bubble devices: (a) the NiFe-first, Au-second structure and (b) the Au-first, NiFe-second structure.

approximately a four-fold reduction in current for the Au-first structure with respect to the Au-second structure.

The problem of dielectric breakdown and resulting short circuit from the current carrying lines to the bubble material is a device problem associated with the amorphous films and not with garnets [unless a metallic layer (Lin and Keefer, 1973; Takahashi et al., 1975) is used for hard bubble suppression]. This problem becomes more severe for small bubble devices because the non-magnetic spacings between the NiFe (and hence the current carrying lines) and the amorphous bubble material decrease approximately in proportion to the bubble diameter. Since the current requirement is about four times less for the Au-first structure than for the NiFe-first structure, the voltage drop is also four times less. The Au-first structure is therefore more advantageous from the point of view of dielectric breakdown as long as the insulating layer between the conductor and the amorphous film has a thickness $t_i > 0.25S$.

Multi-mask-level designs are presently standard in bubble technology, and although such devices have not been fabricated on amorphous films, it is instructive to consider the implications of multi-level designs with amorphous films. The inherent advantage of single-mask-level designs is the fact that there are no critical alignments required. This may well become a severe limitation for bubble devices using features smaller than 1 μm where the amorphous films appear most promising. Additionally, multi-level designs

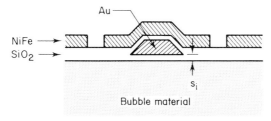

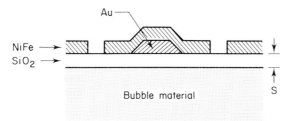

Fig. 32 Alternative structures for multi-level-mask amorphous film bubble devices. (a) The conventional structure in use today in which the conductor level is insulated from the NiFe level. (b) A structure which allows maximum thickness dielectric between the conductor level and the amorphous film.

typically are problematic in the area of the transfer switches where the NiFe must step over the conductor, as shown in Fig. 32a. Obtaining a continuous NiFe element of constant thickness across such a step is a problem in device processing. Furthermore, the fact that the NiFe is not planar causes adverse changes in the coupling of the bubble domain to the NiFe on that element. As a result, that element is more susceptible to propagation failures. Nevertheless, bubble devices are being made today which operate satisfactorily in spite of this "step coverage" problem. If alignment adequate for submicron bubbles can be achieved, the multi-level designs could in fact relax the dielectric breakdown requirements on amorphous films even with conductor-first processing. This is possible since, in multi-level designs, the spacing t_i can in principle be made equal to S with the step in the permalloy equal in height to the full conductor thickness as in Fig. 32b. (Many multi-level switch designs used today do not in fact operate if the conductor is shorted to the NiFe, but certainly any design which could operate as a single-mask-level switch would not be adversely affected by bringing it in contact with the NiFe.) Hence the dielectric breakdown problem in amorphous film bubble devices could find some relief in the use of multi-level transfer switch designs in which the NiFe contacts the conductor.

In addition to the dielectric breakdown problem, the amorphous films put a restriction on the temperatures which can be used during device processing.

This, as discussed earlier, is due to annealing prior to crystallization. This restriction indeed reduces the number of options for device processing, but both dry processes involving evaporation and/or sputtering as well as wet processes involving electroplating have been used to fulfil the requirements. With NiFe evaporated at temperatures less than 200°C, it was necessary to use laminated films to reduce the coercivity and to obtain adequately high magneto-resistive effects. This process was described in detail by Ahn *et al.* (1975). Ahn used lift-off for pattern definition. Powers and Horstman (1974) and Romankiw *et al.* (1974) have previously described processes for bubble device fabrication which used electroplating for the metallization deposition. Electroplating is a low-temperature process and provides the added advantage of precise pattern replication with small lines. The magneto-resistive effect in plated films has been found to be about 2%, whereas values as high as 4% have been reported for permalloy evaporated at high temperature. Thus a 2:1 penalty in detector sensitivity occurs with electroplated permalloy.

8.1 Device performance

Amorphous film bubble devices, operating in a restricted temperature range, have been found to perform well for frequencies of operation from 0.1 Hz to 1 MHz. At high frequency the Gd–Co–Mo amorphous films may in fact have an advantage over other materials in that no hard bubble effects have been observed in the devices operating up to 1 MHz and no added processing such as ion implantation has been necessary for hard bubble suppression.

Figure 33 shows quasi-static margins for H-I bar propagation in a $2\mu m$ bubble size GdCoMo device. The minimum drive of about 25 Oe is typical of $2\mu m$ bubble size garnet and amorphous film devices made on materials with $4\pi M_s \approx 400G$. The full margins at 60 Oe drive are about $\pm 12\%$, also typical of both garnet and amorphous film devices.

Figure 34 shows the increase in minimum drive with increasing frequency for another GdCoMo device (BJL43-3). The minimum drive in this case was determined as that at which approximately 10^6 error-free propagation cycles took place, and is a considerably more stringent requirement than that for the quasi-static margins in Fig. 33. As is expected for materials with finite wall mobility, the minimum drive field increases linearly with frequency. The slope of this line is consistent with a wall mobility of about $2\,m/sec \cdot Oe$ as determined from the device modeling of Almasi and Lin (1976). This value is in good agreement with the data presented earlier on bubble dynamics.

The decrease in margin with the logarithm of the number of bubble steps in H–I bar loops was measured on the amorphous films and found to be

Amorphous Gd–Co Alloys

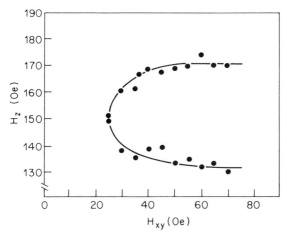

Fig. 33 Quasi-static propagation margins for 2-μm bubbles in H–I bar minor loops on an amorphous film.

less than 0.5 Oe/logarithmic decade on devices with initial margins greater than 25 Oe. A more precise measure of the decrease in margin per step could not be made because temperature variations over the time periods required for such measurements significantly altered the magnetization of the material producing considerable scatter in the data. Devices were, however, run for 10^9 cycles without error. These results indicate that the amorphous films can be made to operate with error rates comparable to those of garnets. It is especially surprising that this is true even though no hard bubble suppression techniques were employed.

In addition to the propagation structure which must operate at high

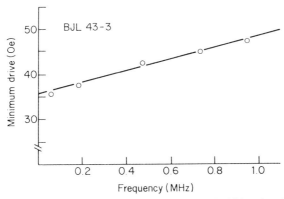

Fig. 34 Minimum in-plane field required propagate bubbles for 10^6 cycles as function of frequency. (From Kryder *et al.*, 1977.)

Fig. 35 A disk replicator.

frequency in bubble devices, disk replicators used for bubble generation and chevron expanders used in detectors require good high-frequency performance of the bubble materials. This is especially true since the average domain wall velocities achieved on these components is typically much higher than those that occur during propagation. These components have also been found to perform very well at high frequency on the amorphous films. Disk generators of the type shown in Fig. 35 have been operated on several amorphous film devices at frequencies up to 1 MHz. Chevron expanders consistently have been found to operate with expansion velocities above 20 m/sec and, as explained before in the section of dynamics, velocities up to 500 m/sec are achieved under certain conditions.

Transfer switches fabricated on amorphous films without severe temperature sensitivity can be expected to operate exactly as they would on garnets. However, due to the strong temperature sensitivity of the 2-μm amorphous films and the fact that 10-mA currents can produce temperature rises of 7.5–

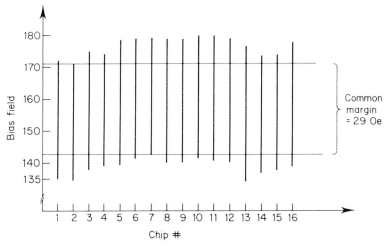

Fig. 36 Propagation margins for 16 chips on a wafer with a drive field of 60 Oe. (From Ahn et al., 1975.)

75°C in the 1 μm × 0.2 μm conductors in these devices depending upon whether they are fabricated on silicon or glass substrates, respectively, most transfer switches operated on amorphous films have been influenced by anomalous temperature effects. Detector signals from chevron expander detectors on amorphous films are, as expected, equivalent to those obtained on garnet films of similar bubble size and magnetization. One example of such signals showed 1 mV/mA sensitivity. Typical margin overlap from different chips is shown in Fig. 36.

8.2 Dielectric breakdown

As evident from the preceding section, bubble domain devices commonly require the use of current-carrying conductors for generating, switching, replicating, and sensing functions. Amorphous rare earth–transition metal films are conductive and consequently electric fields exist between the conductors and the film. If the electric fields are large; these could lead to failure due to breakdown of the insulating films separating the conductors from the storage film. Shorts in the insulator films can also adversely affect device yield.

Dielectric film requirements are dictated by device design factors, specifically conductor geometry and current I. The voltage V across a conductor is determined by Ohm's law, $V = J\rho L$, where J is the current density, ρ is the conductor resistivity, and L is its length. The current density is given by

I/wt_c where w is the conductor width and t_c its thickness. The voltage V develops across the insulator separating the conductor from the storage film. The corresponding electric field in the insulator, E, is given by $E = V/t_i = J\rho L/t_i$, where t_i is the insulator thickness. The quality of an insulator is measured by its ability to support this electric field without failure.

Both yield and reliability of insulator films compatible with amorphous film properties and bubble devices have been investigated by Tan et al. (1978). The principal conclusions of their study are:

(1) Only electric fields smaller than 6×10^5 V/cm provide adequate reliability over approximately 10-year lifetimes for sputtered quartz spacers thicker than 500 Å.
(2) Better than 95% yield is expected at this or smaller fields in bubble devices organized in the major–minor loop configurations employing quartz spacers thicker than 500 Å.

The 6×10^5 V/cm limit corresponds to the threshold above which self-healing breakdown defects are the predominant failure mechanism. This limit is small compared to limits for high quality quartz or the intrinsic limit of defect free quartz. For example, thermally grown or CVD-grown SiO_2 films of the type used in semiconductor devices are rated at two or three times this limit and the intrinsic limit of defect free quartz is as high as 70×10^5 V/cm. The smaller limit for sputtered quartz results from a much higher defect level in films made by this process, presumably due to low film growth and substrate temperature, a necessary constraint for amorphous films.

The yield is high because conductors occupy a small fraction of chip surface in major–minor loop designs. It is expected to be lower in devices which require a larger number of conductors, for example, conductor-access bubble lattice devices. The 500-Å minimum insulator thickness is the minimum thickness investigated by Tan et al. (1978). They emphasized measurements of 500- to 3000-Å thick SiO_2 films. Additional work will be required to assess films thinner than 500 Å both in terms of dielectric integrity and yield.

The consequences of these limits can be assessed by considering the example of a field access device with T–I bar or half-disk propagation elements in which both the conductors and propagation elements are fabricated with only one high-resolution mask (SLM design). The conductor film in SLM field access devices can, in principle, be placed above or below the permalloy propagation elements. However, practical considerations discussed in the preceding section suggest that the conductor be interposed between the propagation elements and the storage film.

We shall consider the geometry sketched in Fig. 31b. In order to deduce the required thickness t_p, of the NiFe in a device, it is noted that the cross-sectional area of the NiFe, $t_p w_2$, can only be reduced in proportion to the flux from the bubble, $4\pi M_s \pi D^2/4$, if saturation of the NiFe is to be avoided. This fact suggests that t_p scales in proportion to DM_s. With 3-μm bubbles and $4\pi M_s \approx 270\,G$, $t_p \approx 4000$ Å in today's devices. Hence it is estimated that $t_p = (5 \times 10^{-4} D)(4\pi M_s)\,(G^{-1})$. The spacing between the center of the NiFe elements and the storage film is typically $D/3$, leaving a total thickness for conductor and insulator, $t_c + t_i$, of $(D/3) - (2.5 \times 10^{-4} D)(4\pi M_s)\,(G^{-1})$. The conductor width w is expected to be $D/2$. Maximizing the current for a given dielectric breakdown field limit E_{max} requires that the total space between the permalloy element and the storage film be equally apportioned to the conductor and insulator, i.e.

$$t_i = t_c = D/[(1.25 \times 10^{-4}\,G^{-1})/D(4\pi M_s)] \tag{26}$$

The maximum current I_{max} which may be put through the conductor is then

$$I_{max} = E_{max} t_i t_c w/\rho L. \tag{27}$$

The longest conductors in field access devices are switch conductors. Assuming a chip dimension of 5 mm square suggests switch conductor lengths L of 5 mm. The geometric factors, the limit $E_{max} = 6 \times 10^5$ V/cm, and a conductor resistivity ρ of $3 \times 10^{-6}\,\Omega$/cm completely establish the maximum switch current I_{max} defined in Eq. (27). Figure 37 depicts the dependence of I_{max} on bubble size (solid curve). This result can be examined in some perspective by comparing the calculated current levels with currents required for various device functions and electromigration constraints.

Published data on field access devices indicate switch currents in the range of 10–100 mA. Even larger currents are required for generating bubbles but these are not of prime concern here since generators employ relatively short conductors and, as can be determined from Eq. (27), develop negligible voltage drops across the conductors. Using switch conductor dimensions in the published literature, the switching currents translate into switch current densities as high as 10^7 A/cm² in Al–Cu conductors.

It is not yet clear whether such high current densities are consistent with attainment of adequate conductor lifetimes from the standpoint of electromigration failure. However, the fact that bubble devices typically require pulsed currents at a relatively low duty cycle, and the possibility that more electromigration failure resistant metallurgies than Al–Cu could be developed in the future, suggest that conductors may perform reliably with current density as high as 10^7 A/cm². This limit is denoted in Fig. 37. (We note that the electromigration considerations alone suggest the need to invent switches which will operate at smaller than heretofore achieved

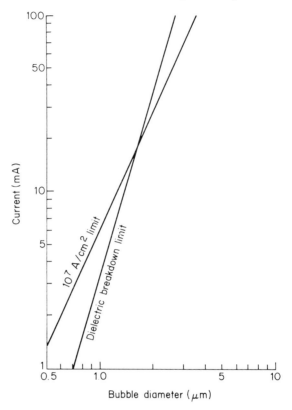

Fig. 37 Maximum electromigration or dielectric breakdown limited current as a function of bubble diameter for a given device geometry.

current levels if micrometer or submicrometer bubble devices, of this or similar type, are to be achieved in the future.)

The dielectric breakdown limit is above the electromigration limit for bubble sizes larger than 1.7 μm. In this range conductor currents will therefore be primarily limited by electromigration considerations. Conversely, dielectric breakdown considerations will limit conductor currents for devices on amorphous films with bubbles smaller than approximately 1.7 μm. This limitation is significant since, as discussed in an earlier section, the amorphous films have been designed to exhibit adequate magnetic properties for bubble sizes below approximately 1 μm. Furthermore, the suggested SLM-conductor first design would provide only 500-Å thick conductors and insulators at a bubble size of approximately 0.5 μm and zero permalloy element to storage film spacing at bubble sizes below approximately 0.25 μm.

Increasing these limits for T–I-bar or half-disk devices will require one or a combination of solutions: attaining insulator films with superior dielectric breakdown strength to that of sputtered quartz, designing switches with conductor lengths shorter than 5 mm, use of multilevel metallurgies with the associated requirement of precise mask registrations at submicrometer dimensions, and inventing switches with lower currents.

9 SUMMARY AND CONCLUSIONS

In this paper we have reviewed the properties of amorphous Gd–Co alloys from the standpoint of magnetic bubble applications. We have neither attempted to include the properties of other amorphous rare-earth–transition metal alloys, nor the amorphous transition metal–metalloid systems which are also of technological interest. We have restricted our attention to the storage media in magnetic bubble technology. Our review suggests that the preparation of amorphous films is reasonably well understood and could be translated into practical fabrication. The desirability of the degree of control of the sputtering parameters, the preparation of targets, substrates, etc. are known. What is currently lacking in amorphous films is the temperature insensitivity requisite for bubble technology using large bubbles. This is particularly apparent for bubble diameters larger than $1\,\mu$m. Other than the temperature sensitivity, the remaining materials' properties are found to be adequate for the conventional field access devices such as the T- and I-bar structures. In 1-μm or smaller bubble diameters the temperature sensitivity improves and these materials become increasingly competitive in comparison to the garnets. Roughly below $0.4\,\mu$m the amorphous materials are, of course, currently favored, as the garnets cannot go down to bubbles smaller than this diameter and the hexa-ferrites have still to be developed.

Our understanding of electromigration and dielectric failure in T- and I-bar device configurations suggests that for bubbles less than $1\,\mu$m the amorphous materials may run into problems associated with their metallic properties. This recognition along with the inadequate temperature sensitivity of the large bubble materials could lead one to conclude that the utility of amorphous materials is limited. We believe this conclusion would be erroneous. Extrapolation of T- and I-bar behavior to small bubbles in garnets would also lead one to be pessimistic about bubble technology at the half micrometer level. This pessimism is based on the lithography problem associated with T- and I-bar devices for half-micrometer bubbles and, more basically, the ability to switch magnetization in a structure (the I bar) whose width is comparable to the domain wall width. An obvious way out is to design structures that are not limited by the state of the art lithography or domain wall width. An example is the contiguous disc device structure or some other pattern which is larger than the bubble and is field accessed. A rough calculation of the type described here for T- and I-bar devices suggests that electromigration and dielectric breakdown would not be serious gating factors for amorphous films using notions associated with contiguous disc type patterns. However, before amorphous films can be considered for contiguous disc devices, ion implantability and the existence of charged walls of the kind present in garnets have to be demonstrated. Recent experimental evidence obtained in this laboratory suggests that this can indeed be accomplished.

The metallic nature of the amorphous Gd–Co films has invariably been viewed as being problematic. It seems to us that this metallic nature could usefully be used to conduct current through the storage layer to locally control bubble motion. Recently experimental evidence for the interaction of magnetic bubbles with electric currents (domain drag effect) has been obtained in amorphous Gd–Co–Mo alloys (DeLuca *et al.*, 1978). This observation could provide the impetus towards a search for practical applications of this phenomena.

When bubble technology moves to bubbles smaller than one micron the amorphous bubble materials are likely to become increasingly important. The physics of these materials is in its infancy, as is the whole area of magnetism in amorphous materials. The origin of magnetic anisotropy and its relationship to structure and preparation conditions remains to be unequivocally established. Control of magnetization, its temperature dependence, the effect of alloying, etc. have received relatively scant attention. The utility of thousand-angstrom bubbles, which has been demonstrated in these materials (Chaudhari and Herd, 1976), is unexplored.

REFERENCES

Ahn, K. Y., Chang, T. H. P., Hatzakis, M., Kryder, M. H., and Luhn, H. (1975). *IEEE Trans. Magn.* **MAG-11**, 1142.
Ali, A., Grundy, P. J., and Stephens, G. A. (1976). *J. Phys. D.* **9**, L69.
Almasi, G. S., and Lin, Y. S. (1976). *IEEE Trans. Magn.* **MAG-12**, 160.
Argyle, B. E., Gambino, R. J., and Ahn, K. Y. (1974). *AIP Conf. Proc.* No. 24, 564.
Axe, J. D., Passell, L., and Tsuei, C. C. (1975). *AIP Conf. Proc.* No. 24, 119.
Bajorek, C. H., and Kobliska, R. J. (1976). *IBM J. Res. Dev.* **20**, 271.
Bobeck, A. H., and Schovill, H. E. D. (1971). *Sci. Am.* **224**, 78.
Bobeck, A. H., Danylchuk, I., Rossol, F. C., and Stross, W. (1973). *IEEE Trans. Magn.* **MAG-9**, 474.
Bonyhard, P. I., and Smith, J. L. (1976). *IEEE Trans. Magn.* **MAG-12**, 614.
Brunsch, A., and Schneider, J. (1977). *IEEE Trans. Magn.* **MAG-13**, 1606.
Burilla, C. T., Bekebrede, W. L., and Smith, A. B. (1976). *AIP Conf. Proc.* No. 34, 340.

Burilla, C. T., Bekebrede, W. R., Kestigian, M., and Smith, A. B. (1977). *Conf. Magn. Magn. Mater., 23rd, Minneapolis.* Pap. ZP-6.
Cargill, G. S., III (1974). *AIP Conf. Proc.* No. 18, 631.
Cargill, G. S., III (1977). *Conf. Magn. Magn. Mater., 23rd, Minneapolis.* Pap. 2P-07.
Cargill, G. S., III, and Kirkpatrick, E. S. (1976). *AIP Conf. Proc.* No. 31, 339.
Cargill, G. S., III, and Mizoguchi, T. (1978). *J. Appl. Phys.* **49**, 1753.
Chaudhari, P., and Cronemeyer, D. C. (1976). *AIP Conf. Proc.* No. 29, 113.
Chaudhari, P., and Herd, S. R. (1976). *IBM J. Res. Dev.* **20**, 2.
Chaudhari, P., and Turnbull, D. (1978). *Science (Washington, D.C.)* **199**, 11.
Chaudhari, P., Cuomo, J. J., and Gambino, R. J. (1973a). *IBM J. Res. Dev.* **17**, 66.
Chaudhari, P., Cuomo, J. J., and Gambino, R. J. (1973b). *Appl. Phys. Lett.* **22**, 337.
Chaudhari, P., Cuomo, J. J., Gambino, R. J., Kirkpatrick, S., and Tao, L. J. (1975). *AIP Conf. Proc.* No. 24, 562.
Cochrane, R. W., and Cargill, G. S., III (1974). *Phys. Rev.* **32**, 476.
Cuomo, J. J., and Gambino, R. J. (1975). *J. Vac. Sci. Technol.* **12**, 79.
Cuomo, J. J., and Gambino, R. J. (1977). *J. Vac. Sci. Technol.* **14**, 152.
Cuomo, J. J., Sadagopan, V., De Luca, J., Chaudhari, P., and Rosenberg, R. (1972). *Appl. Phys. Lett.* **12**, 581.
de Gennes, P. (1962). *J. Phys. Radium* **23**, 630.
DeLuca, J. C., Gambino, R. J., and Malozemoff, A. P. (1978). *IEEE Trans. Magn.* **MAG-14**, 500.
Dillon, J. F., Jr. (1963). *In* "Magnetism" (G. T. Rado and H. Suhl, eds.), Vol. 3, p. 450. Academic Press, New York.
Dirks, A. G., and Leamy, H. J. (1977). *Conf. Magn. Magn. Mater., 23rd, Minneapolis* Pap. 2P-1.
Durand, J. (1985). *J. Phys. F* (in press).
Esho, S., and Fujiwara, S. (1976). *AIP Conf. Proc.* No. 34, 331.
Forester, D. W. (1976). *AIP Conf. Proc.* No. 31, 384.
Frait, Z., Nagy, I., and Tarnoczi, T. (1976). *Phys. Lett. A* **55**, 429.
Gambino, R. J., and Cuomo, J. J. (1978). *J. Vac. Sci. Technol.* **15**, 296.
Gambino, R. J., Ziegler, J. F., and Cuomo, J. J. (1974a). *J. Appl. Phys.* **24**, 99.
Gambino, R. J., Chaudhari, P., and Cuomo, J. J. (1974b). *AIP Conf. Proc.* No. 18, 578.
Gangulee, A., and Kobliska, R. J. (1975). *J. Appl. Phys.* (in press).
Gangulee, A., and Taylor, R. C. (1977). *Conf. Magn. Magn. Mater., 23rd, Minneapolis* Pap. 2P-11.
Geiss, R. (1976). *In* "Developments in Electron Microscopy and Analysis" (J. Venables, ed.), p. 61. Academic Press, New York.
George, P. K., Hughes, A. J., and Archer, J. L. (1974). *IEEE Trans. Magn.* **MAG-10**, 821.
Gill, H. S., Chen, M., and Judy, J. H. (1977). *Conf. Magn. Magn. Mater., 23rd, Minneapolis* Pap. 2P-3.
Graczyk, J. F. (1976). *AIP Conf. Proc.* No. 34, 343.
Graczyk, J. F. (1977). *Conf. Magn. Magn. Mater., 23rd, Minneapolis* Pap. 2P-2.
Hafner, D., and Humphrey, F. B. (1977). *Appl. Phys. Lett.* **30**, 303.
Hagedorn, F. B. (1974). *J. Appl. Phys.* **45**, 3129.
Hasegawa, R., and Taylor, R. C. (1975). *J. Appl. Phys.* **46**, 3606.
Hasegawa, R., Argyle, B. E., and Tao, L. J. (1974a). *AIP Conf. Proc.* No. 24, 110.
Hasegawa, R., Gambino, R. J., Cuomo, J. J., and Ziegler, J. F. (1974b). *J. Appl. Phys.* **45**, 4037.

Heiman, N., and Lee, U. (1974). *Phys. Rev. Lett.* **33**, 778.
Heiman, N., and Lee, K. (1975a). *Phys. Lett. A* **55A**, 297.
Heiman, N., and Lee, K. (1975b). *AIP Conf. Proc.* No. 24, 108.
Heiman, N., and Lee, K. (1976). *AIP Conf. Proc.* No. 34, 319.
Heiman, N., Onton, A., Kyser, D. F., Lee, K., and Guarnieri, C. R. (1974). *AIP Conf. Proc.* No. 24, 573.
Heiman, N., Lee, K., and Potter, R. I. (1976). *AIP Conf. Proc.* No. 29, 130.
Herd, S. (1977). *Conf. Magn. Magn. Mater., 23rd, Minneapolis* Pap. 2P-4.
Herd, S., and Chaudhari, P. (1974). *Phys. Status Solidi A* **26**, 627.
Hoffman, H., and Höpfl, R. (1976). *Int. Conf. Magn. Bubbles, Eindhoven, Conf. Program*, 51.
Josephs, R. M., and Stein, B. F. (1975). *AIP Conf. Proc.* No. 24, 598.
Katayama, T., Hasegawa, K., Kawanishi, K., and Tsushima, T. (1977). *Conf. Magn. Magn. Mater., 23rd, Minneapolis* P. Z-10.
Kim, S., Henderson, D., and Chaudhari, P. (1977). *Thin Solid Films* **47**, 155.
Kirshnan, R., Suran, G., Sztew, J., Jouve, H., and Meyer, R. (1977). *Conf. Magn. Magn. Mater., 23rd, Minneapolis* Pap. 2P-12.
Kobliska, R. J., and Gangulee, A. (1975). *AIP Conf. Proc.* No. 24, 567.
Kobliska, R. J., Ruf, R., and Cuomo, J. J. (1975). *AIP Conf. Proc.* No. 24, 570.
Kobliska, R. J., Gangulee, A., Cox, D. E. and Bajorek, C. H. (1977). *IEEE Trans. Magn.* **MAG-13**, 1767.
Kryder, M. H., and Hu, H. L. (1974). *AIP Conf. Proc.* No. 18, 213.
Kryder, M. H., Ahn, K., Almasi, G. S., Keefe, G. E. and Powers, J. V. (1974). *IEEE Trans. Magn.* **MAG-10**, 825.
Kryder, M. H., Ahn, K. Y., and Powers, J. V. (1976). *IEEE Trans. Magn.* **MAG-11**, 1145.
Kryder, M. H., Tao, L. J. and Wilts, C. H. (1977). *IEEE Trans. Magn.* **MAG-13**,1626.
Lin, S. C. M. (1969). *J. Appl. Phys.* **40**, 2175.
Lin, Y. S., and Keefe, G. E. (1973). *Appl. Phys. Lett.* **22**, 603.
Lin, Y. S., Almasi, G. S., and Keefe, G. E. (1977). *IEEE Trans. Magn.* **MAG-13**, 1744.
McGoll, J. R., Murphy, D., Cargill, G. S., III, and Mizoguchi, T. (1976). *AIP Conf. Proc.* No. 29, 172.
McGuire, T. R., Taylor, R. C., and Gambino, R. J. (1976). *AIP Conf. Proc.* No. 34, 346.
McGuire, T. R., Gambino, R. J., and Taylor, R. C. (1977a). *J. Appl. Phys.* **48**, 2965.
McGuire, T. R., Gambino, R. J., and Taylor, R. C. (1977b). *IEEE Trans. Magn.* **MAG-13**, 1598.
Meyer, R., Jouve, H., and Rebouillat, J. P. (1975). *IEEE Trans. Magn.* **MAG-11**, 1335.
Minamura, Y., Imamura, N., and Kobayashi, T. (1976). *IEEE Trans. Magn.* **MAG-12**, 779.
Minkiewicz, V. J., Albert, P. A., Potter, R. I., and Guarnieri, C. R. (1976). *AIP Conf. Proc.* No. 29, 107.
Mizoguchi, T. (1976). *AIP Conf. Proc.* No. 34, 286.
Mizoguchi, T., McGuire, T. R., Gambino, R. J., and Kirkpatrick, S. (1977a). *Physica B + C (Amsterdam)* **86–88B** + **C**, 783.
Mizoguchi, T., Gambino, R. J., Hammer, W. N., and Cuomo, J. J. (1977b). *IEEE Trans. Magn.* **MAG-13**, 1618.
Mizoguchi, T., Malozemoff, A. P., and Cox, D. E. (1977c). *Conf. Magn. Magn. Mater., 23rd, Minneapolis* Pap. ZP-8.

Nandra, S. S., and Grundy, P. J. (1971). *J. Phys. F* **1**, 207.
Okamoto, K., Sirokuwa, T., Matsushita, S., and Sakurai, Y. (1976). *IEEE Trans. Magn.* **MAG-10**, 799.
Onton, A., and Lee, K. (1976). *AIP Conf. Proc.* No. 34, 328.
Orehotsky, J., and Schroeder, K. (1972). *J. Appl. Phys.* **43**, 2413.
Potter, R. I., Minkiewicz, V. J., Lee, K., and Albert, P. A. (1976). *AIP Conf. Proc.* No. 29, 76.
Raj, K., Budnick, J. I., Alben, R., Chi, G. C., and Cargill, G. S., III. (1976). *AIP Conf. Proc.* No. 31, 390.
Rhyne, J. J., Pickart, S. J., and Alperin, H. A. (1974). *AIP Conf. Proc.* No. 18, 563.
Romankiw, L. T., Krongelb, W. S., Castellani, E. E., Pfeifer, A. T., Stoeber, B. J., and Olsen, J. D. (1974). *IEEE Trans. Magn.* **MAG-10**, 828.
Schneider, J., and Brunsch, A. *Conf. Magn. Magn. Mater., 23rd, Minneapolis* Pap. ZP-5.
Schneider, J. W. (1975). *IBM J. Res. Dev.* **19**, 587.
Sharon, T. E., and Tsuei, C. C. (1972). *Phys. Rev. B* **5**, 1047.
Slonczewski, J. (1973). *J. Appl. Phys.* **44**, 1759.
Slonczewski, J., Malozemoff, A. P., and Voegeli, D. (1973). *AIP Conf. Proc.* No. 10, 458.
Sunago, K., Matsushita, S., and Sakurai, Y. (1976). *IEEE Trans. Magn.* **MAG-12**, 776.
Suran, G., Krisnan, R., Jouve, H., and Meyer, R. (1977). *IEEE Trans. Magn.* **MAG-13**, 1532.
Takahashi, M., Nishida, N., and Kobayashi, T. (1975). *J. Phys. Soc. Jpn.* **34**, 1416.
Tan, S. I., Ainslie, N., Ahn, K. Y., Cox, D., and Bajorek, C. (1978). *IEEE Trans. Magn.* **MAG-14**.
Taylor, R. C. (1976a). *AIP Conf. Proc.* No. 29, 190.
Taylor, R. C. (1976b). *J. Appl. Phys.* **47**, 1164.
Taylor, R. C., and Gangulee, A. (1976). *J. Appl. Phys.* **47**, 4666.
Taylor, R. C., McGuire, T. R., Coey, J. M. D., and Gangulee, A. (1985). *J. Appl. Phys.* (in press).
Tsuei, C. C., and Lillienthal, H. (1976). *Phys. Rev. B* **13**, 4899.
Vella-Coleiro, G. P., and Tabor, W. J. (1972). *Appl. Phys. Lett.* **21**, 7.
Venturini, E. L., Richards, P. M., Borders, J. A., and Fer Nisse, E. P. (1976). *AIP Conf. Proc.* No. 29, 116.
Voegeli, O., Calhoun, B. A., Rosier, L. L., and Slonczewski, J. C. (1975). *AIP Conf. Proc.* No. 24, 617.
Vossen, J. L. (1971). *J. Vac. Sci. Technol.* **8**, 512.
Wagner, C. N. J., Heiman, N., Huang, T. C., Onton, A., and Parrish, W. (1976). *AIP Conf. Proc.* No. 29, 188.
Zwingman, R., Wilson, W. C., and Bourne, H. C. (1976). *AIP Conf. Proc.* No. 34, 334.

3 Garnet Films for High Bubble Velocities and High Bubble Mobilities

D. J. BREED and U. ENZ

*Philips Research Laboratories
Eindhoven, The Netherlands*

1 Introduction 92
   1.1 Material parameters 92
   1.2 Dynamical properties of domain walls and bubbles in garnets 93
2 Garnet films with orthorhombic anisotropy 97
   2.1 Film preparation 97
   2.2 Magnetic properties of the $(Gd, Y)_3(Fe, Mn, Ga)_5O_{12}$ Films 103
   2.3 Static properties of bubbles 113
   2.4 Dynamic properties 120
3 Bismuth-containing garnet films with high uniaxial anisotropy and low damping 126
   3.1 Introduction 126
   3.2 Growth of bismuth-containing films 127
   3.3 Anisotropy 127
   3.4 Dynamic properties of the bismuth-containing garnet films 128
   3.5 Temperature dependence of bubble parameters . . . 129
4 Device implications 130
   4.1 Current access devices on garnet films with orthorhombic anisotropy 130
   4.2 Device potential of high-mobility bismuth-substituted garnets for bubble diameters $<1\,\mu m$ 133
5 Conclusion 133
   References 134

1 INTRODUCTION

In this introduction the basic material parameters that are of importance for bubble materials with uniaxial as well as orthorhombic anisotropy are given and the present understanding concerning the dynamical properties will be outlined. In Section 2, the results obtained in materials with orthorhombic anisotropy are discussed. The Bi-containing bubble materials with high uniaxial anisotropy energy and low damping are treated in Section 3. The device implications of both types of materials are discussed in Section 4.

1.1 Material parameters

Bubble materials are characterized by a number of parameters which govern their static as well as their dynamical properties. When devising a bubble material, these parameters must be controlled in such a way as to optimize the performance of the bubble device and to meet the specifications of a particular application. The final solution will often be a compromise between conflicting demands. An excellent treatment of this subject can be found in the book of Eschenfelder (1980).

Bubble materials are used in the form of thin films of thickness h having a uniaxial anisotropy which favours an easy axis perpendicular to the film plane. Static material parameters include the saturation magnetization M_s, the uniaxial and cubic anisotropy constants K_u and K_1, respectively, the cubic magnetostriction constants λ_{111} and λ_{100}, and the exchange stiffness parameter A.

A primary condition for uniaxial bubble materials is

$$Q = K_u/2\pi M_s^2 > 1, \qquad (1)$$

where Q is the "quality factor" of the material and K_u the value of the uniaxial anisotropy energy. If the condition of Eq. (1) is fulfilled, the general direction of the magnetization is oriented normal to the film plane, opposing the restoring force of the demagnetizing field. If $Q > 1$, then the internal structure of the domain wall is dominated by the local anisotropy and by the exchange energy, and the influence of magnetostatic stray fields can be taken into account as a correction. This guarantees also that the domain wall width is independent of the velocity of the domain wall. A Q-factor of the order of 2 to 4 turns out to be appropriate for bubble device materials, the former figure applying to bubbles sizes $\leq 1\,\mu$m, the latter to larger bubbles.

Garnet Films

A physical parameter of importance is the specific wall energy σ_w:

$$\sigma_w = 4(AK_u)^{1/2}, \tag{2}$$

where A is the exchange constant.

The specific wall energy and the saturation magnetization jointly determine the material length parameter

$$l = \sigma_w/4\pi M_s^2, \tag{3}$$

a fundamental material property which directly governs the size of domains. In a film of thickness $h = 4l$, the diameter of a bubble will be equal to $8l$. This specific value of h is the optimum film thickness in the sense that it yields the minimum bubble size. It also yields a large bias field margin of $\pm 17\%$ relative to the mean bias field of $0.28 \times 4\pi M_s$ (Thiele, 1971). Some deviation from the optimum film thickness is, however, allowed.

A film thickness near to $h = 9l$ is often used in actual devices. The relative bias field margins of these "thick films" are then smaller, about 13% relative to the mean bias field of $0.46 \times 4\pi M_s$, but the absolute field margins are larger. Generally speaking, the bubble size is always of the same order of magnitude as the film thickness. This implies that a film carrying micrometer-sized bubbles must have a thickness of the order of 1 μm and must therefore be preparable as an epitaxial or sputtered film on a non-magnetic substrate. The material lengths of known materials range from $l \simeq 1.5 \times 10^{-6}$ cm for $BaFe_{12}O_{19}$, a hexagonal oxide (Ferroxdure), over $l = 2 \times 10^{-5}$ to 10^{-3} cm for substituted garnets to $l \simeq 10^{-3}$ to 10^{-2} cm for various orthoferrites. At present, only garnets are used in actual devices, mainly because of the high flexibility of garnets to adjust the material parameters by chemical substitutions.

1.2 Dynamical properties of domain walls and bubbles in garnets

Apart from the static requirements, a bubble material should also meet certain requirements related to the dynamics of bubble motion. The motion of bubbles is hampered by at least two effects: wall pinning as expressed by the static coercive force, and wall damping due to a viscous drag opposing the motion of the walls. Both effects will contribute to a continuous dissipation of energy associated with a moving bubble and should therefore be minimized. Coercivity is strongly correlated with the occurrence of imperfections in the magnetic films; it can be reduced by increasing crystal perfection. The coercive force in bubble films as expressed by $H_c/4\pi M_s$ should be lower than 0.05. Values of 0.02 are now commonly used. Wall damping in garnets is found to be caused mainly by intra-atomic electronic relaxations inherent to some anisotropic rare earth ions. In this chapter we will show that by avoiding these relaxing ions it is possible to realize high-mobility materials even for submicrometer bubbles. In the latter films Bi ions have been substituted in order to increase the uniaxial anisotropy. The mobilities realized range from 100 m/sec·Oe for the films with orthorhombic anisotrophy to 300 m/sec·Oe for the Bi-containing films with uniaxial anisotropy.

A further interesting property of Bi-containing bubble materials is their extremely high specific Faraday rotation. Although this parameter is not of direct importance to the functioning of the bubble device, it is, for testing purposes, very useful.

Rapid progress has been made in the study of the dynamical properties of domain walls and bubbles in recent years. One of the results concerns the insight that there is an upper limit to the velocity of a domain wall or a bubble, the peak velocity, that cannot be overcome by higher driving fields; other results concern the internal structure of bubbles, e.g. the creation of Bloch lines as a consequence of the bubble motion. Furthermore, new experiments have led to the elimination of the large discrepancies which apparently existed between damping parameters obtained from ferromagnetic resonance (FMR) experiments and those obtained from domain wall mobility studies. Here, we would like to trace briefly some of these results. Detailed reports have been published by de Leeuw *et al.* (1980) and by Malozemoff and Slonczewski (1979).

The physical basis of domain wall motion is the gyroscopic precession of the electronic spin (or the magnetization vector) as described by the Landau–Lifshitz equation: the magnetization vector responds orthogonally to a torque acting on it by an external or internal field. Domain wall motion can be understood qualitatively by starting on this physical basis. Let us consider the simplest case, a Bloch wall in an infinite medium having uniaxial anisotropy. The preferential direction is parallel to the z-axis of a coordinate system x, y, z. The direction of the magnetization is defined by the polar angle θ and the azimuthal angle ϕ. Inside the resting wall (Fig. 1a) the magnetization direction changes in a continuous way from the $-z$ direction to the $+z$ direction, while the magnetization vector always remains parallel to the yz plane, the plane of the wall. The structure can therefore be described with the aid of the angle $\theta = \theta(x)$ alone.

In the presence of an external field H_z the wall moves, i.e. the magnetization at a certain point in the material rotates from the $-z$ to the $+z$ direction during the passage of the wall. In view of the gyroscopic properties of the magnetization, this can only occur in the presence of a magnetic field H_x perpendicular to the wall. After the application of the external field H_z the

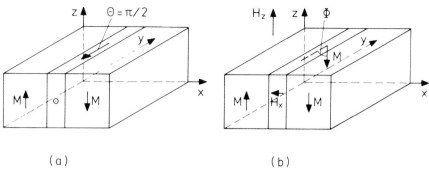

Fig. 1 Straight-planar Bloch wall in a uniaxial material (a) at rest and (b) in motion under the influence of an applied external field H_z. A precession of the magnetization vector around the demagnetization field occurs inside the wall, causing the wall to move to the right.

magnetization first precesses around the z-axis, thereby building up a component M_x perpendicular to the plane of the wall. This component, in turn, provides the demagnetization field $H_x = -4\pi M_x$. In a steady state, when the work delivered by the external field just compensates the losses introduced by the damping, the components H_z, M_x, and thus ϕ are all constant. As the precession frequency is γH_x, the wall velocity is proportional to H_x. This steady-state situation is shown in Fig. 1b.

From the above qualitative picture it becomes clear that a wall has a finite maximum velocity. The reason is that the demagnetization field H_x is necessarily finite ($H_x \leq 4\pi M_s$), implying a finite precession frequency and thus a finite maximum velocity. It also becomes clear that an additional static in-plane field will increase the precessional frequency and therefore the peak velocity. This effect has indeed been observed by de Leeuw (1977). In a film of $(Y, La)_3(Fe, Ga)_5O_{12}$ he measured the wall velocity as a function of the drive field at a constant in-plane field (Fig. 2). A small and nearly field independent velocity of about 10 m/sec was observed in zero in-plane field (curve A). In an in-plane field of 187 Oe, a much larger wall velocity was observed. At low driving field, $H < 2$ Oe, a linear region was found, which can be described by $v = \mu H$, the peak velocity being about 300 m/sec. The wall velocity drops abruptly for $H > 2$ Oe and finally increases again (curve B). The wall mobility is of the order of 100 m/sec·Oe.

Another way to increase the peak velocity also follows directly from the physical base outlined above: it is the use of a material with orthorhombic anisotropy. In this case, the additional static in-plane field needed is an anisotropy field. Rossol (1971) found indeed very high velocities in $YFeO_3$, an orthoferrite. The influence of an in-plane anisotropy field has been treated in detail by Hagedorn (1971). Stacy et al. (1976) and Breed et al. (1977)

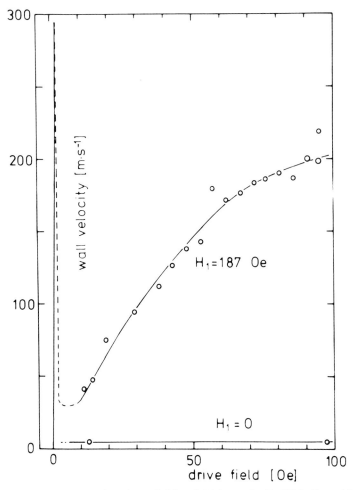

Fig. 2 Wall velocity as a function of drive field in an LaGa YIG film with and without in-plane field H_1. Curve A, $H_1 = 0$; curve B, $H_1 = 187$ Oe. (After de Leeuw, 1977.)

have shown that orthorhombic anisotropy can also be introduced in (cubic) garnet, leading indeed to a large increase in maximum domain-wall velocity. The orthorhombic anisotropy was a result of a combination of stress, introduced by liquid phase epitaxy (LPE) growth on (110)-oriented GGG substrates and suitable magneto-elastic constants. It is important to note that this can be achieved, by Mn^{3+} substitution, without introducing rare-earth ions. A large part of the present work is devoted to this subject, which will be treated in detail.

2 GARNET FILMS WITH ORTHORHOMBIC ANISOTROPY

The first garnet films with orthorhombic anisotropy and high domain-wall velocity (Stacy et al., 1976) were $(Eu, Lu)_3(Fe, Al)_5O_{12}$ films. In these films the damping and coercivity were quite high. An improvement was the discovery that the substitution of Mn^{3+} ions strongly increases the stress-induced anisotropy, leaving damping and coercivity low (Breed et al., 1977). In this section our attention will be focussed mainly on the latter films with nominal composition $(Gd, Y)_3(Fe, Mn, Ga)_5O_{12}$, the magnetic properties and growth conditions of which have been published recently (Breed et al., 1983).

2.1 Film preparation

2.1.1 Growth on (110) oriented substrates

Orthorhombic anisotropy is obtained by growing garnet films on (110) substrate surfaces. The films we report on were all grown using the normal liquid phase epitaxy (LPE) technique, with horizontal dipping and with rotation at 120 rpm in the reverse mode (5 sec). For garnets (110) is a habit face, which means that growth may occur by virtue of a two-dimensional nucleation or of a spiral growth mechanism. Figure 3 shows the experimental

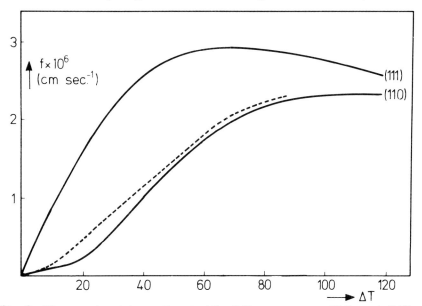

Fig. 3 The experimental growth rate (f) of films grown on (111)- and (110)-oriented substrates as a function of supercooling (ΔT). The dotted line indicates the growth rate of the hillocks.

Fig. 4 Hillocks on the (110)-grown films.

growth rate f as a function of supercooling ΔT for the orientations $\{110\}$ and $\{111\}$. Inspection of (110)-oriented films shows that hillocks emerge from the surface (Fig. 4). Hillocks with heights up to $2\,\mu$m and base cross-section up to $500\,\mu$m occur. Between the hillocks the grown film is smooth. The growth of the hillocks and the growth of the smooth film in between is assumed to occur by virtue of two different mechanisms. The hillocks probably grow by means of the spiral growth mechanism, described by Burton *et al.* (1951), in which a screw dislocation with its Burgers vector ending on the surface acts as a continuous step source. Since substrate treatment appears to have an influence on the density of hillocks, surface irregularities and impurities from the melt also seem to act as step sources. The growth of the smooth film is due to the two-dimensional nucleation mechanism, which is effective only when the supercooling exceeds some critical value. Below this critical supercooling, the growth rate f should be zero. Slight misorientation of the substrates, which were oriented with an accuracy of $0.1°$, may explain the low but measurable growth rate of the (110) oriented films at low supercoolings (Fig. 3). As a result of the misorientation, the substrate surface is stepped and these steps will move, leading to growth. [From the experimental curve the critical supercooling for growth on (110) is about 15°C.] The dotted curve in Fig. 3 gives the growth rate of the hillocks when it is assumed that hillocks originate on the substrate surface. As is to be expected from the theory of spiral growth, the relation between f and ΔT is a quadratic one and the curve passes through the origin. Hillocks are observed up to very high supercoolings, although their heights above the smooth film become smaller when ΔT increases. For a device one needs smooth films of constant thickness. We found that LPE growth on deliberately misoriented substrates yields perfectly smooth films. This is explained by the presence of a sufficiently stepped surface which gives rise to a growth rate higher than that of the hillocks. Experiments on the misorientation versus the growth rate indicated that a misorientation of $1°$ is sufficient to get rid of the hillocks up to $\Delta T = 60°C$. The growth of garnet films on (110)-oriented surfaces and the occurrence of hillocks as function of misorientation and supercooling has been discussed in detail by van Erk et al. (1980). Besides the introduction of orthorhombic anisotropy, growth on (110) has two other advantages:

(1) The growth rate is lower than on (111), even when the sample is misoriented. This permits an easier and more accurate control of thickness, which is especially important for thin films.
(2) Due to the low growth rate, volume diffusion in the melt is less important and transient effects hardly occur. Ferromagnetic resonance experiments revealed that the composition of the (110) grown films is more uniform than that of the (111) films.

2.1.2 Influence of manganese on growth conditions

The Mn/Fe ratio in the film is dependent on the Mn/Fe ratio in the melt. Figure 5 shows the molar ratios of film and melt for $(Gd, Y)_3(Fe, Mn)_5O_{12}$ films grown with a supercooling of about $25°C$. It appeared that at the same supercooling the Mn/Fe ratio in the film is hardly affected by the substitution of Ga. This is also shown for two films in Fig. 5.

The influence of manganese on the growth conditions is surprisingly different from the influence of other substituents like Y, La, Ga, and Fe. Van Erk (1979) has shown that the saturation temperature of a system increases if the concentration of substituting components in the melt is increased. However, as shown in Fig. 6a, the saturation temperature strongly decreases when manganese oxide is added to the solution. Furthermore, addition of manganese oxide causes a decrease of the slope of the growth rate versus supercooling curve. In Fig. 6b the growth rate per degree supercooling as a function of the Mn/Fe ratio in the solution is presented. These two effects cause the strong decrease in film growth rate at constant growth temperature when manganese oxide is added to the solution (Breed et al., 1977).

The maximum manganese concentration in the films is determined by

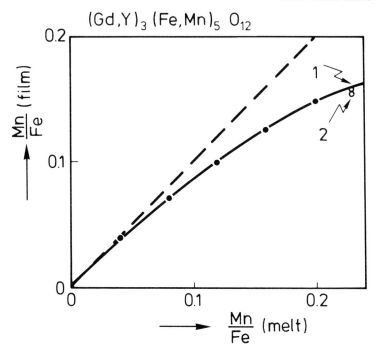

Fig. 5 Molecular ratio of Mn/Fe in the film versus Mn/Fe in the melt. The black points are for films without Ga. Films 1 and 2 are Ga-containing films from melts 1 and 2, respectively (see Table 1).

occurrence of second phase in the melt for Mn/Fe ratios in the melt larger than 0.25. The second phase appeared to be a crystalline material consisting of Fe, Mn, Pb, and O.

2.1.3 Melt Compositions

The melt compositions for films with orthorhombic anisotropy for 2 and 1 μm bubbles are shown in Table 1. In this case $MnCO_3$ has been used to introduce manganese in the melt. Mn_2O_3 and MnO_2 were also used. It appeared that this has no influence on the final result. It was argued (Voermans et al., 1979) that manganese is substituted as Mn^{3+}. This is also consistent with electron microprobe analysis.

For device films slightly misoriented from (110), GGG substrates have been used. The misorientation was 1° as indicated in Fig. 14. The direction of the misorientation is very important as will be seen later on (Section 2.2.5).

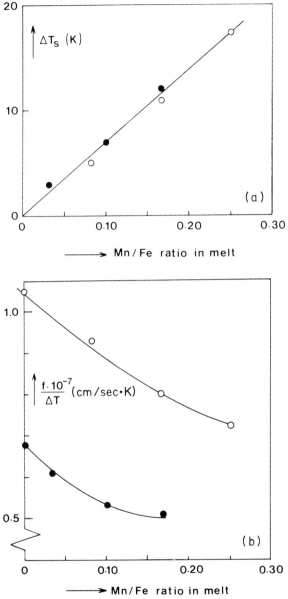

Fig. 6 Influence of manganese on growth conditions. (a) The decrease of the saturation temperature due to manganese for two melts with a different concentration of garnet forming oxides is given. Open circles, $T_s = 949 - \Delta T_s$; solid dots, $T_s = 901 - \Delta T_s$. (b) The growth rate, which is approximately a linear function of the supercooling, is shown as a function of the Mn/Fe ratio in the melt.

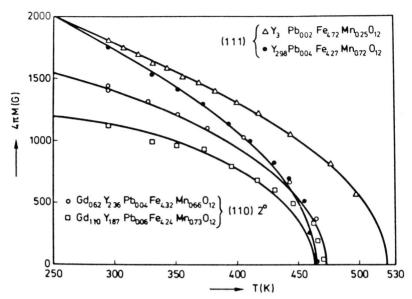

Fig. 7 Magnetization versus temperature of films with a different manganese and gadolinium concentrations; (111)- and (110)-oriented films were used. The (110)-oriented films were grown with a misorientation of 2°. The drawn curves were calculated using the molecular field approximation.

Table 1 Melt composition and properties.

Parameter	Melt 1[c]	Melt 2[c]
PbO[a]	12.0	12.0
B_2O_3[a]	0.8	0.8
Fe_2O_3[a]	1.576	1.588
$MnCO_3$[a]	0.725	0.731
Ga_2O_3[a]	0.0762	0.0367
Y_2O_3[a]	0.0376	0.0439
Gd_2O_3[a]	0.0311	0.0373
T_s[b]	935°C	—
T_g[b]	910°C	933°C
$f(110)$[b]	0.4 μm/min	0.7 μm/min
$\emptyset$[b]	2 μm	1 μm

[a] The melt compositions are given in mole.
[b] T_s, saturation temperature; T_g, dip temperature; f, growth velocity of the film; $\emptyset$, bubble diameter under device conditions.
[c] Composition of the films as determined by electron microprobe analysis:
$Gd_{1.35}Y_{1.64}Pb_{0.02}Fe_{3.93}Mn_{0.62}Ga_{0.43}O_{12}$ (melt 1);
$Gd_{1.39}Y_{1.62}Pb_{0.03}Fe_{4.13}Mn_{0.64}Ga_{0.20}O_{12}$ (melt 2).

2.2 Magnetic properties of the $(Gd, Y)_3(Fe, Mn, Ga)_5O_{12}$ films

The Mn^{3+} ion is a magnetic ion which interacts antiferromagnetically with neighbouring Fe^{3+} ions. The Mn^{3+} ion affects the magnetization, Curie temperature, and gyromagnetic ratio. Furthermore, the concentration of Mn^{3+} ions in combination with the mismatch between film and substrate determines the anisotropy in the films. The misorientation introduced to obtain perfect films affects strongly the preferential direction of magnetization.

2.2.1 Magnetization versus temperature

The molecular field theory, usually failing in predicting ordering temperatures from basic principles, appears to be very powerful to describe the saturation magnetization versus temperature in the RE iron garnet system if use is made of phenomenological parameters as is shown by Dionne (1979) and Röschmann and Hansen (1981). Dionne introduced the phenomenological constants N_{dd}, N_{aa}, N_{ad}, N_{dc}, and N_{ac}. These constants describe the interaction between the magnetic ions in the different magnetic sublattices with d = tetrahedral, a = octahedral, and c = dodecahedral site. Magnetic dilution was also taken into account.

The general chemical formula for our composition is

$$\{Gd_xY_{3-x}\}[Fe_{2-y-z_1}Mn_yGa_{z_1}](Fe_{3-z_2}Ga_{z_2})O_{12} \qquad (4)$$

The brackets { } in this case are for the dodecahedral ions, [ ] for the octahedral ions, and ( ) for the tetrahedral ions. The manganese ions decrease the mean interaction between the octahedral and tetrahedral sublattices and decrease the octahedral sublattice magnetization. Furthermore, it appeared that in the temperature range between room temperature and the Curie temperature the fit between calculated and measured magnetization versus temperature curves can be improved by modifying some of the constants introduced by Dionne (1979). Following Röschmann and Hansen (1981), we use

$$N_{ad} = 80 \qquad N_{aa} = 42.6 \qquad N_{dd} = -17.6. \qquad (5)$$

For the influence of Gd we take

$$N_{dc} = 6.6 \qquad N_{ac} = -3.44, \qquad (6)$$

where N_{dc} is somewhat different from the value used by Dionne. The influence of Mn has been taken into account by introducing

$$N^*_{ad} = (1 - p + \alpha p)N_{ad} \qquad \text{and} \qquad S^* = \tfrac{5}{2}(1 - \tfrac{1}{5}p), \qquad (7)$$

where $p = y/(2 - z_1)$, which is the manganese fraction of octahedral magnetic

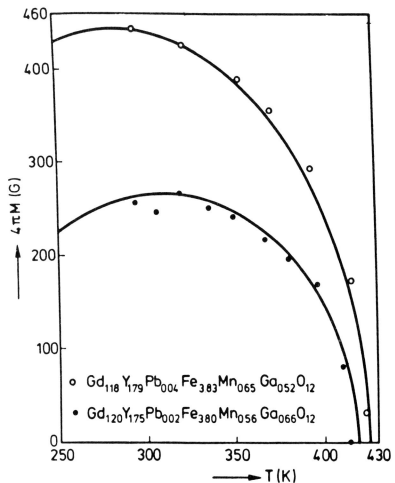

Fig. 8 Influence of Ga on the magnetization versus temperature curves. These films are (111)-oriented films. The drawn curves were calculated using the molecular field approximation taking into account a small concentration of Ga ions on the octahedral sites (Röschmann, 1980).

ions, and α is the ratio of the octahedral–tetrahedral Mn–Fe and Fe–Fe exchange interaction. The correction for the spin value has to be applied only for the octahedral sublattice. [We assume in this case that $S(Mn^{3+}) = 2$.] A good fit with the experimental data can be obtained with

$$\alpha = 0.80. \qquad (8)$$

We assumed that a small part of the Ga ions occupy the octahedral sublattice as is shown by Röschmann (1980). In Figs. 7 and 8 it is shown that with the above modifications of the molecular field approximation of Dionne, the magnetization can be predicted quite well.

It appeared that the influence of the manganese ions on the Curie temperature compares very well with the influence of the Ga ions. The lowering of the Curie temperature can be approximated by

$$\Delta T_c = 120(y + z), \qquad (9)$$

where $z = z_1 + z_2$. From this lowering of the Curie temperature, the change in exchange parameter can be estimated (Slonczewski et al., 1974):

$$A(T, y, z) = A_0(T)\{1 - \Delta T_c/(559 - T)\}, \qquad (10)$$

where 559 is the Curie temperature of YIG and A_0 the exchange constant of YIG.

2.2.2 Gyromagnetic ratio

The g-value of Fe^{3+} and Gd^{3+} is quite accurately 2, and the g value of Mn^{3+} is not known. Since the saturation magnetization of $(Gd, Y)_3$-$(Fe, Mn, Ga)_5O_{12}$ is a result of a difference between much larger sublattice magnetizations, small deviations from $g = 2$ in one sublattice may have considerable consequences for the g value of the total systems. It can be shown that

$$g \simeq 2\left[1 - \frac{M_a}{M_s} \cdot \frac{(4 - 2g_{Mn})y}{10 - (5 - 2g_{Mn})y}\right], \qquad (11)$$

where M_a is the magnetization of the octahedral sublattice, M_s the saturation magnetization, and g_{Mn} the g value of the Mn^{3+} ions. From a comparison of different films, where the gyromagnetic ratio was measured by means of FMR, we estimated

$$g_{Mn} = 1.956 \pm 0.014. \qquad (12)$$

2.2.3 Lattice parameter

From electron microprobe analysis, it appears that Pt and Pb are present in the film as contaminants. The Pt concentration (0.002–0.003 Pt per formula unit) can be neglected, but the Pb concentration (0.02 to 0.07 Pb per formula unit) contributes significantly to the lattice parameter. If the symbols for the elements indicate the concentration per formula unit, then the lattice parameter can be described quite well by

$$a \ (\text{Å}) = 12.376 + 0.0317 \, \text{Gd} + C_1 \text{Pb}$$
$$+ C_2(\text{Gd} + \text{Y} + \text{Pb} - 3) + C_3 \text{Mn} + C_4 \text{Ga}. \qquad (13)$$

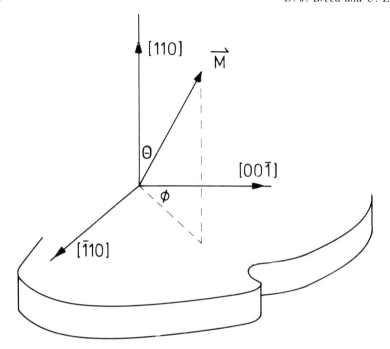

Fig. 9 Stress-induced anisotropy in a (110)-oriented film: $F=(K_u + K_i \sin^2 \phi) \sin^2 \theta$; $M_\perp \to K_u$, $K_u + K_i > 0$; $K_u = -\tfrac{3}{4}\sigma(\lambda_{100} + \lambda_{111})$; $K_i = +\tfrac{3}{4}\sigma(\lambda_{100} - \lambda_{111})$.

The first two terms in this equation are an interpolation between the lattice parameters of YIG (12.376 Å) and GdIG (12.471 Å); $C_1 = 0.08 \pm 0.006$ (W. de Roode, personal communication), $C_2 = 0.20$, $C_3 = -0.007 \pm 0.004$, and $C_4 = -0.017$.

We assume that Gd and Pb are always substituted on the dodecahedral sites. C_2 is due to excess Y substituted on the octahedral sites. This constant has been calculated using the ionic radius of Y^{3+} in an octahedral environment (0.892 Å) and calculating the influence of this on the lattice parameter (Strocka et al., 1978). C_3 we determined experimentally from a large number of manganese containing films, and C_4 is taken from Strocka et al. (1978).

2.2.4 Anisotropy and magnetostriction

The introduced non-cubic magnetic anisotropy in a (110)-oriented film is orthorhombic and can be written in first approximation as (see Fig. 9)

$$F = (K_u + K_i \sin^2 \phi) \sin^2 \theta, \tag{14}$$

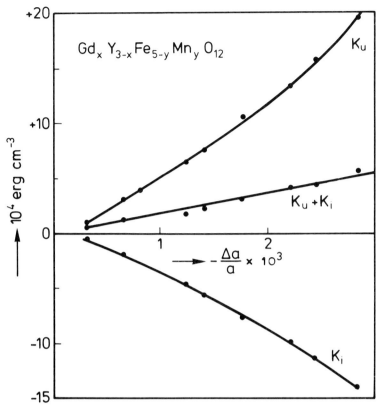

Fig. 10 Anisotropy as a function of the misfit between film and substrate. In these films, the manganese concentration was nearly constant; $\Delta a/a$ was changed by changing the Gd content from 0.28 to 1.22 per formula unit.

where K_i describes the anisotropy in the plane of the film. In the $(Gd, Y)_3$ $(Fe, Mn, Ga)_5 O_{12}$ films, the anisotropy is a result of the combination of high magnetostriction constants and stress in the film due to a mismatch between the lattice parameters of film and substrate. In these films the growth-induced anisotropy can be neglected as can be concluded from Fig. 10, where the anisotropy is shown as function of the mismatch between film and substrate. In this case the concentration of manganese is high (between 0.63 and 0.78 per formula unit) and the mismatch was varied by changing the gadolinium concentration from 0.28 to 1.22 per formula unit. It is found that the residual anisotropy for small values of $\Delta a/a$, which should be the growth-induced anisotropy, is negligible.

The magnetic anisotropy as function of the manganese concentration is shown in Fig. 11. The mismatch between film and substrate is nearly

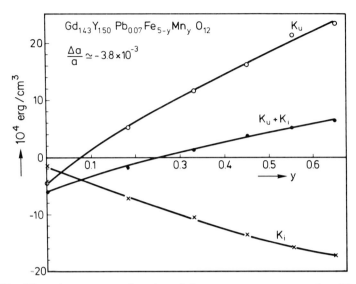

Fig. 11 The anisotropy as a function of the manganese concentration. The misfit in these films was nearly constant.

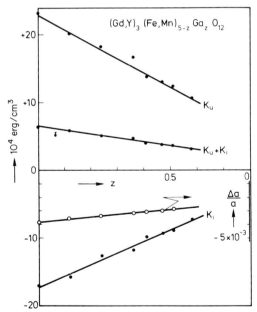

Fig. 12 Anisotropy as a function of Ga concentration. The Gd/Y ratio was about 0.92 and $y \simeq 0.62$. The decrease in lattice parameter due to the Ga substitution is given on the right-hand side.

constant. These anisotropies are sufficiently high for small bubble devices, but one has to take into account the fact that the dilution with Ga reduces the anisotropy as is shown in Fig. 12. The manganese concentration and the Gd concentration are nearly constant, about 0.6 per formula unit and 1.4 per formula unit, respectively. The lattice parameters slightly decrease, due to the Ga substitution [see Eq. (13)]. This decrease in lattice parameter explains only 50% of the decrease in anisotropy. The other part is due to the change in the exchange constant, which appears to have a similar effect on the stress-induced anisotropy as was already observed for the growth-induced anisotropy in uniaxial garnet films (Eschenfelder, 1977).

The anisotropies in these films are given by (Breed et al., 1978–1979):

$$K_u = -\tfrac{3}{4}\sigma(\lambda_{100} + \lambda_{111}), \quad K_u + K_i = -\tfrac{3}{2}\sigma\lambda_{111}, \quad K_i = +\tfrac{3}{4}\sigma(\lambda_{100} - \lambda_{111}), \tag{15}$$

where

$$\sigma = 2.67 \times 10^{12}\, \Delta a/a \quad \text{dyne/cm}^2, \tag{16}$$

Δa is the mismatch between film and substrate, and $\Delta a < 0$ if the film is in compression.

Phenomenologically, the anisotropy and magnetostriction can be described very well in the following way. For the magnetostriction of the film we write

$$\lambda = [(1 - z/2)\lambda^\circ + y\lambda^{Mn}]A(T, y, z)/A_0(T), \tag{17}$$

where λ° is the magnetostriction of YIG.

The term $(1 - z/2)$ describes the influence of magnetic dilution by the Ga ions on the magnetostriction constant of YIG (White, 1973), and $y\lambda^{Mn}$ is the contribution to the magnetostriction of the manganese ions, not taking into account the effect that manganese also reduces the exchange interaction. The term $A(T, y, z)/A_0(T)$ describes the reduction of the magnetostriction due to a reduction of the exchange interaction. We adopted this term phenomenologically in the same way as Eschenfelder (1977) did for the growth induced anisotropy. At room temperature the exchange constant of YIG is given by (Gerhardstein et al., 1978) $A_0 = 3.75 \times 10^{-7}$ erg/cm^2. Using Eq. (10), the room temperature magnetostriction of the film is then given by

$$\lambda = [(1 - z/2)\lambda^\circ + y\lambda^{Mn}][1 - 0.45(y + z)]. \tag{18}$$

The experimental anisotropies can be predicted using Eqs. (15), (16), and (18) with the phenomenological parameters

$$\lambda^\circ_{111} = -3.9 \times 10^{-6}, \quad \lambda^\circ_{100} = -1.9 \times 10^{-6},$$
$$\lambda^{Mn}_{111} = +14.9 \times 10^{-6}, \quad \lambda^{Mn}_{100} = +60.2 \times 10^{-6}. \tag{19}$$

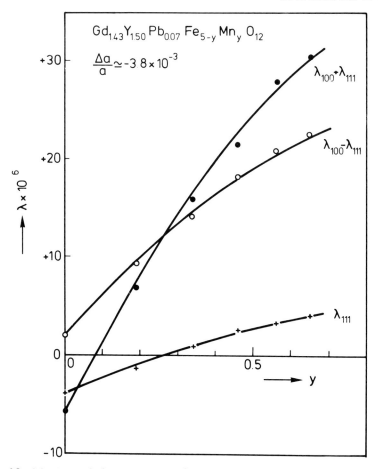

Fig. 13 Magnetostriction constants of manganese-containing films. The drawn curves were calculated, using Eqs. (18)–(19). The experimental points were obtained by using the experimental data of Fig. 11 and Eq. (15).

In Fig. 13 we have plotted the magnetostriction constants calculated from Fig. 11. The drawn curves were calculated using equations (18) and (19). The absolute values of λ° are larger than observed in YIG crystals by Hansen (1974): $\lambda_{100} = -1.25 \times 10^{-6}$ and $\lambda_{111} = -2.73 \times 10^{-6}$. The discrepancy may be due to a slight contamination by Fe^{2+} in the YIG single crystals. The values of L^{Mn} agree quite well with the data of Dionne and Goodenough (1972) for manganese in YIG single crystals. From their data at $y = 0.26$, we estimate $\lambda_{111}^{Mn} = +14.3 \times 10^{-6}$, $\lambda_{100}^{Mn} = +76 \times 10^{-6}$ to be compared with 14.9×10^{-6} and 60.2×10^{-6}, respectively.

2.2.5 Influence of misorientation on magnetic properties

If the film normal is misoriented with respect to the [110] direction, then the easy axis is not well defined, as shown by Pierce (1972). The direction of the easy axis is no longer parallel to the film normal or to the [110] direction and the angle between the easy axis and the film normal is generally a strong function of the misorientation.

It can easily be shown (Yatsenko *et al.*, 1979) that

$$\beta/\delta = \frac{-\frac{3}{2}\sigma(\lambda_{100} - \lambda_{111})}{\frac{3}{2}\sigma(\lambda_{100} + \lambda_{111}) + 4\pi M^2} \quad (20)$$

if the misorientation is in the ($\bar{1}$10) face (see Fig. 14), and

$$\beta/\delta = \frac{3\sigma(\lambda_{100} - \lambda_{111})}{3\sigma\lambda_{111} + 4\pi M^2} \quad (21)$$

if the misorientation is in the (00$\bar{1}$) face, where δ is the angle between the [110] axis and film normal and β the angle between magnetization and film normal.

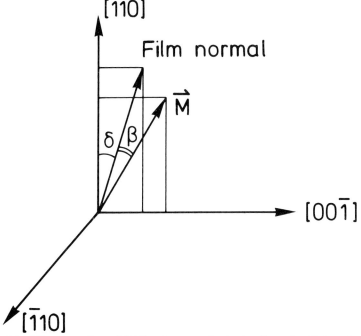

Fig. 14 Misorientation of (110) substrates. In device films, $\delta \simeq 1°$ (see Section 2, 2, 1). The direction of the easy axis makes an angle β with the film normal (see Section 2, 2, 5).

The term $4\pi M^2$ is due to the demagnetizing field not taken into account by Yatsenko et al. (1979). From Fig. 13 and these equations, it can be seen that $\beta/\delta \simeq -1$ and $\beta/\delta \gg 1$ in the two cases, respectively. In the first case the magnetization is more or less fixed to the [110] direction, whereas in the second case the misorientation of the magnetization is much larger than the misorientation of the film. Therefore we have misoriented the film in the ($\bar{1}$10) face, as indicated in Fig. 14.

2.2.6 Materials tailoring

The important material parameters for tailoring the magnetic properties in the manganese containing films with orthorhombic anisotropy [Eq. (14)] are the quality factors

$$Q_1 = K'_u/2\pi M_s^2 \quad \text{and} \quad Q_2 = -K_i/2\pi M_s^2 \qquad (22)$$

where

$$K'_u = K_u + K_i$$

and the materials length l given by Eq. (3). The exchange parameter in l is determined by the Curie temperature [Eq. (10)]. The parameter Q_1 is the same as Q in Eq. (1).

In Table 2 the calculated and experimental values of $\Delta a/a$, $4\pi M$, T_c, K_u, K_i, $K_u + K_i$, and g are compared. The calculated values were obtained as follows: $\Delta a/a$ from Eq. (13), $4\pi M$ and T_c as described in Section (2.2.1) K_u

Table 2 Calculated and measured film parameters.[a]

Parameter	Film 1[b]		Film 2[b]		Film 3[b]	
	Calc.[a]	Exp.	Calc.[a]	Exp.	Calc.[a]	Exp.
$-(\Delta a/a) \times 10^3$	2.2	2.6	3.1	2.66	2.4	2.7
$4\pi M$ (G)	462	452	702	734	144	141
T_c (K)	433	438	452	463	443	443
$K_u \times 10^{-4}$ (erg/cm^3)	11.6	11.6	14.6	15.5	5.41	5.07
$K_i \times 10^{-4}$ (erg/cm^3)	-8.2	-8.4	-10.4	-10.9	-4.38	-4.65
$(K_u + K_i) \times 10^{-4}$ (erg/cm^3)	3.4	3.5	4.1	4.4	1.03	0.58
g	2.096	2.067	2.064	2.078	2.143	2.236

[a] The calculated values were obtained as described in Section 2.2. The anisotropies were calculated using the experimental values of $\Delta a/a$.
[b] Film 1: $Gd_{1.35}Y_{1.64}Pb_{0.02}Fe_{3.93}Mn_{0.62}Ga_{0.43}O_{12}$ (melt 1). Film 2: $Gd_{1.39}Y_{1.62}Pb_{0.03}Fe_{4.13}Mn_{0.64}Ga_{0.20}O_{12}$ (melt 2). Film 3: $Gd_{1.42}Y_{1.53}Pb_{0.05}Fe_{4.03}Mn_{0.28}Ga_{0.68}O_{12}$.

and K_i from Eqs. (15) to (19). The agreement is sufficient for material tailoring.

2.3 Static properties of bubbles

In films with strong in-plane anisotropy, the mean domain-wall magnetization will be directed along the axis of medium anisotropy energy, the [$\bar{1}10$] axis in the (110)-oriented $(Gd, Y)_3(Fe, Mn, Ga)_3O_{12}$ films. The domain wall of a magnetic bubble is therefore partly a Néel wall and partly a Bloch wall. The two simplest configurations are shown in Fig. 15a. These configurations are not necessarily the configurations with the lowest domain-wall energy. In the Néel segments part of the domain wall magnetization is directed in the direction opposite to the surface stray field. In Fig. 15b a bubble domain-wall configuration with one horizontal and two vertical Bloch lines is shown, where the contribution of surface stray fields to the domain wall energy has

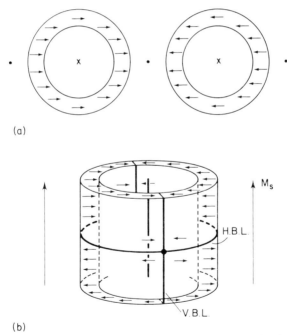

Fig. 15 (a) The two different A-type bubbles. The wall magnetization is in the direction of the axis of medium anisotropy energy. (b) A possible structure for the B-type bubbles with one horizontal (H.B.L.) and two vertical Bloch lines (V.B.L.). The net wall magnetization in the wall is parallel to the surface stray field if the surface stray field is parallel to the axis of medium anisotropy energy.

been minimized. These structures are similar to the structures suggested by Konishi *et al.* (1981).

Schlömann (1976) calculated the critical value of Q_2 below which value it becomes favourable to have a horizontal Bloch line in a static one-dimensional domain wall perpendicular to the axis of medium anisotropy energy,

$$Q_2^{cr} = 1 + (\ln 2)^2 \pi M_s^2 h^2 / 2A. \tag{23}$$

For a film under device conditions ($h = 9l, Q_1 \simeq 3$), this critical value of Q_2 is about 120, which is larger than the Q_2 values that have been realized in the $(Gd, Y)_3(Fe, Mn, Ga)_5O_{12}$ films. From this point of view it is likely to have horizontal Bloch-line segments in the Néel parts of the bubble domain wall.

An in-plane field in the direction of the axis of medium anisotropy has a strong influence on the static properties of the bubbles as can be understood from Fig. 15 (see Section 2.3.2). The in-plane field has also a strong influence on the strip pattern of the $(Gd, Y)_3(Fe, Mn, Ga)_5O_{12}$ films (Breed and

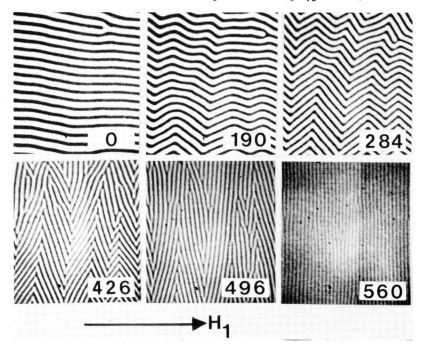

Fig. 16 Domain pattern of a (110)-oriented film in zero bias field; H_1 is parallel to the axis of medium anisotropy energy. The numbers are the values of H_1 in oersteds. The arrow indicates the direction of H_1.

Voermans, 1980). This latter effect is due to the large magneto-striction constants of these films.

2.3.1 Strip domain under influence of in-plane fields

In the manganese-containing films with orthorhombic anisotropy, we observed the anomalous behaviour that in high in-plane fields parallel to the axis of medium anisotropy energy, the strip domains orient themselves perpendicular to the in-plane field, indicating that in this situation Néel walls are more favourable than Bloch walls. This rearrangement of magnetic domains under the influence of an in-plane field can be observed in two ways: (1) the change of a strip domain pattern in an increasing in-plane field; (2) the nucleation of strip domains in a fixed in-plane field. Figure 16 shows the change of the strip domain pattern, according to (1). It is clear from this figure that in large in-plane fields the Néel walls perpendicular to the in-plane field H_1 become energetically favourable. The direction of the nucleated strip domains changes from parallel to H_1 at smaller values of H_1 to perpendicular at larger values of H_1, as shown in Fig. 17. The first and the last photographs in this figure show clearly the influence of H_1 on the direction of the nucleated strips. In the other photographs, where the bias

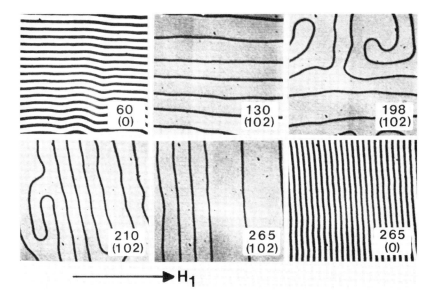

Fig. 17 Development of strip domains in decreasing bias field. The numbers in parentheses are the values in oersteds to which the bias field has been lowered. The other numbers are the fixed in-plane fields.

field is just below the nucleation field, the in-plane field range is shown where the direction of the strips changes from parallel to perpendicular to H_1.

The behaviour shown in Figs. 16 and 17 can be explained partly by taking into account the influence of strains near the domain walls on the domain wall energy. Strains parallel to the domain wall cannot occur due to the rigidity of the substrate. However, strains perpendicular to the domain wall can occur in very small areas near the domain walls and are not suppressed by the substrate. It was shown (Breed and Voermans, 1980), that there is a critical in-plane field H_1^{cr} where Néel walls becomes more favourable than Bloch walls:

$$H_1^{cr} = (2K_u'/M_s) \frac{2\pi M_s^2 - \frac{9}{8} E\lambda_{111}(\lambda_{100} + \lambda_{111})}{2\pi M_s^2 + \frac{9}{8} E\lambda_{111}(\lambda_{100} + \lambda_{111})}, \qquad (24)$$

where E is the modulus of elasticity. The influence of H_1 is due to the resulting tilt of the magnetization. The positive demagnetizing energy of the Néel wall decreases more rapidly with H_1 than the negative strain-induced contribution to the domain wall energy. At H_1^{cr}, both competitive energy contributions are in equilibrium.

2.3.2 Influence of in-plane fields on the static properties of bubbles

The bubbles without Bloch lines we call A-type bubbles (Breed et al., 1982; Breed and de Geus, 1983) (see Fig. 15a). These bubbles are generated in the presence of an in-plane field during bubble generation. If no in-plane field is present several types of bubbles can be generated. If Q_2 is not very large as is the case in film 1 and 2 the main part of the generated bubbles are with Bloch lines. These bubbles were called B-type bubbles. In Fig. 15b an example of such a bubble was shown. It appeared that A-type and B-type bubbles have different static properties: the collapse fields are different and the dependence of collapse field and bubble diameter on the in-plane field is very different. The collapse field, measured with zero in-plane field, of the A-type bubbles decreased with respect to the B-type bubbles from 213 to 208 Oe in film 1 and from 504 to 496 Oe in film 2. This difference in collapse field gave the opportunity to distinguish both types of bubbles.

The B-type bubbles were rather unstable under influence of in-plane fields. In-plane fields of 100–200 Oe were sufficient to change a B-type bubble to an A-type bubble. The different behaviour of A-type and B-type bubbles under influence of in-plane fields is shown in Fig. 18 for the 1-μm bubble film. In this case the collapse field was measured as a function of the in-plane field. Whereas near zero for the in-plane field the collapse field of the A-type bubbles strongly depends on the in-plane field, $dH_{col}/dH_1 \simeq 0.14$,

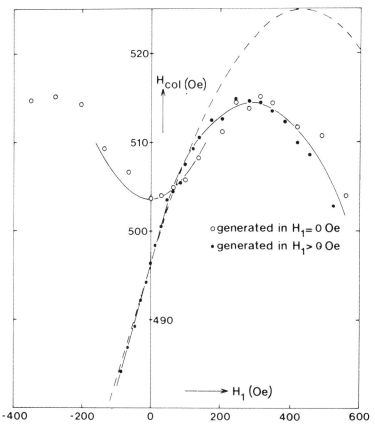

Fig. 18 The collapse field as a function of the in-plane field for film 2. The black dots are results for the A-type bubbles and the open circles for the B-type bubbles. Note that for in-plane fields larger than 100 Oe, transition occurs from B-type to A-type bubbles. The broken line is the theoretical curve according to Eq. (27). The solid lines represents experimental results. For $H_1 = 0$, the collapse field is 496 Oe.

the collapse field of the B-type bubbles does not depend on the in-plane field, $dH_{col}/dH_1 \simeq 0$.

Also, the bubble diameter of the A-type bubbles strongly depends on the in-plane field. Bubbles generated in a positive in-plane field increase in size in positive in-plane fields and decrease in size in negative in-plane fields. In fact, two different A-type bubbles can be distinguished depending on the sign of H_1 during bubble generation.

Under certain conditions it appears to be possible to have both types of bubbles simultaneously in a film, as shown in Fig. 19. For reasons of clarity a film with larger bubble diameter ($\phi \simeq 4\,\mu$m) was chosen. Figure 19 shows

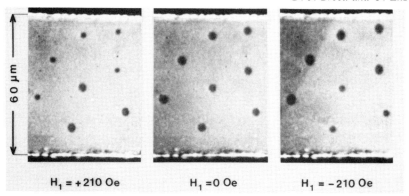

Fig. 19 Bubbles of type A with opposite wall magnetization in in-plane fields parallel to the axis of medium anisotropy energy.

that small and large A-type bubbles occur simultaneously under influence of an in-plane field and that the bubble size reverses when the sign of the in-plane field is reversed.

The bubble diameter as a function of the in-plane field was analysed in the large bubble diameter film 3. Results for the A-type bubbles in this film are shown in Fig. 20. The in-plane field dependence was measured at three values of the bias field. The sign of the in-plane field is taken positive if the in-plane field has the same direction as the wall magnetization.

Using a similar approximation as proposed by Dorleijn et al. (1973), it can be shown that the materials length for the A-type bubbles as a function of the in-plane field H_1 is given by

$$l(H_1) = l(0)\left\{\cos\theta_0 - \left(\frac{\pi}{2} - \theta_0\right)\sin\theta_0\right\}\bigg/\cos^2\theta_0, \quad (25)$$

where θ_0 is the tilt of the magnetization due to the in-plane field H_1 given by

$$\sin\theta_0 = M_s H_1/2K_u. \quad (26)$$

The collapse field as a function of H_1 is then given by

$$H_{col}(H_1) = 4\pi M_s \cos\theta_0[1 - \{3l(H_1)/4h\}^{1/2}]^2. \quad (27)$$

This result is also plotted in Fig. 18, fitting $4\pi M_s$ in such a way that the collapse field at $H_1 = 0$ is 496 Oe. It can be seen that the observed dependence on H_1 is qualitatively well reproduced. The agreement at low values of H_1 is very good.

Along the same lines the bubble diameter as a function of the in-plane

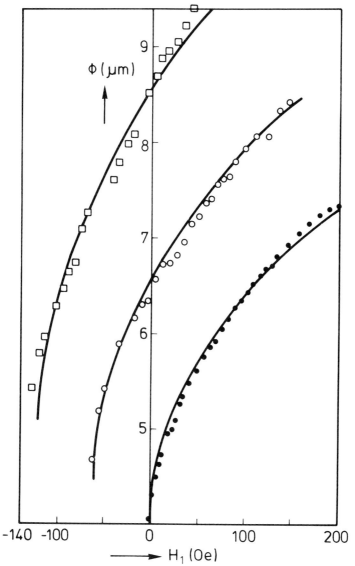

Fig. 20 Bubble diameter as a function of the in-plane field for A-type bubbles in film 3 (Table 2). The experiments were done at three different bias fields: 32.5, 29.5, and 26.6 Oe. The drawn curves were calculated fitting $4\pi M_s$ at $H_1 = 0$. In this case the sample was aligned carefully in order to have no component of H_1 in the direction of the easy axis.

field was calculated using the equation of Callen and Josephs (1971) with the modification

$$4\pi M_\perp = 4\pi M_s \cos\theta_0. \tag{28}$$

So the bubble diameter ϕ is given by

$$\frac{\phi(H_1)}{h} = \frac{\left[1 - \frac{H}{4\pi M_\perp} - \frac{3l(H_1)}{4h}\right] + \left[\left(1 - \frac{H}{4\pi M_\perp} - \frac{3l(H_1)}{4h}\right)^2 - \frac{3l(H_1)H}{4\pi M_\perp h}\right]^{1/2}}{(3H/8\pi M_\perp)} \tag{29}$$

It is shown in Fig. 20 that this equation reproduces the experimental curves very well if $4\pi M_s$ is adjusted.

2.4 Dynamic properties (Breed et al., 1983)

The anisotropy in the (110)-oriented films is given by Eq. (14). We neglect in this equation the slight misorientation of the film. For the manganese-containing films, $K_i < 0$ and $K'_u = K_u + K_i < K_u$. With the quality factors defined in Eq. (22), the maximum velocity of domain walls is given by (Schlömann, 1976)

$$v^a_{max} = 2\gamma(2\pi A Q_1)^{1/2}\left[\left(1 + \frac{1}{Q_1} + \frac{Q_2}{Q_1}\right)^{1/2} - 1\right], \tag{30}$$

$$v^b_{max} = 2\gamma(2\pi A Q_1)^{1/2}\left|\left(1 + \frac{Q_2}{Q_1}\right)^{1/2} - \left(1 + \frac{1}{Q_1}\right)^{1/2}\right|. \tag{31}$$

Here, v^a_{max} corresponds to the case where the wall is parallel to the medium axis (the in-plane [$\bar{1}10$] direction) and v^b_{max} to the case where wall and medium axis are perpendicular. The predicted anisotropy in the velocity is small if Q_2 is large. Experimentally, however, lower maximum velocities and larger anisotropies in the maximum velocity have been observed.

2.4.1 Straight domain-wall experiments

Straight domain-wall experiments have been done in equipment built by de Leeuw (1973). The advantage of these experiments is the simple wall configuration. However, in connection with the available magnetic gradient field to stabilize a straight wall (Hagedorn, 1970), special samples with low magnetization and small thickness have to be prepared. A result for such a film is shown in Fig. 21. The parameters of this film are the thickness

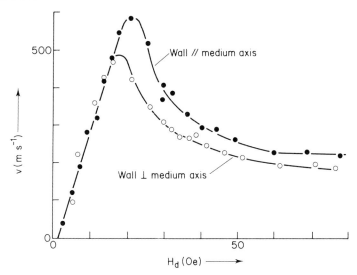

Fig. 21 Straight domain-wall velocity of a $(Gd, Y)_3(Fe, Mn, Ga)_5O_{12}$ film. The mobility is $14\,\text{m/sec·Oe}$.

$h = 0.77\,\mu\text{m}$, $4\pi M_s = 185\,\text{G}$, $Q_1 = 11.7$, $Q_2 = 32.8$, $\gamma = 1.93 \times 10^7/\text{sec·Oe}$ and $A = 1.52 \times 10^{-7}\,\text{erg/cm}^2$. The critical value of Q_2 that can be calculated for this film from Eq. (23) is 7.4, so one should not expect horizontal Bloch lines in this film. In spite of this, there is a clear discrepancy between theory and experiment $v_{\max}^a = 600\,\text{m/sec}$, $v_{\max}^b = 460\,\text{m/sec}$, to be compared with the calculated values from Eqs. (30) and (31) of $1280\,\text{m/sec}$ and $1200\,\text{m/sec}$, respectively. An important result of Fig. 21 is the fact that the mobility is isotropic, so one can conclude that bubble propagation is isotropic in these anisotropic films as long as $v < v_{\max}$.

The observations that $v_{\max}$ is smaller and that the anisotropy in $v_{\max}$ is stronger than predicted are quite general (see, e.g., Breed et al., 1978). The reason for this discrepancy is not clear at present and may be due to a mechanism for domain-wall instability different from theory, such as the creation of horizontal Bloch lines.

2.4.2 Bubble translation experiments

The translational bubble velocities were measured in the way described by Vella-Coleiro and Tabor (1972). No bias field compensation was used, and two pairs of parallel conductors gave a field gradient of $6.9\,\text{Oe}/\mu\text{m}$. The minimum pulse length was about 10 ns, and the maximum current pulse about 1 A (pulse generator EH 1421/1121).

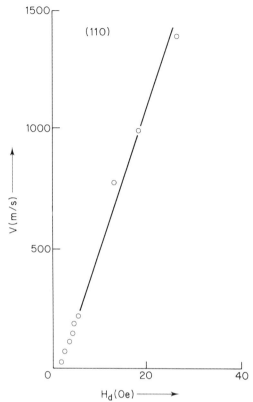

Fig. 22 Bubble translation velocity as a function of drive field. The drive field H_d is the field gradient × bubble radius. Composition: $(Gd, Y)_3(Fe, Mn, Ga)_5O_{12}$.

As the light source we used a mercury lamp or a 7-nsec laser pulse. In the latter way we avoided misinterpretation due to overshoot effects (Breed et al., 1980). In Fig. 22, a result is shown for film 3 with bubbles of about 8 μm diameter. The material parameters of this film are $h = 2.73$ μm, $4\pi M = 141$ G, $Q_1 = 7.3$, $Q_2 = 58.9$, $\gamma = 1.97 \times 10^{-7}$/Oe·sec, and $A = 2.2 \times 10^{-7}$ erg/cm². In this case the bubbles moved in the direction of the gradient field. The maximum velocity is larger than about 1400 m/sec and the mobility is 55 m/sec·Oe. It was also shown that the bubble mobility in this film was isotropic (Breed et al., 1980). The maximum velocity could not be measured in this film in connection with the maximum bubble translation that could be observed in the shortest pulse time. The extremely high velocity of 1400 m/sec is a result of the very high value of Q_2 [v_{max} from Eqs. (30) and (31) is about 2600 m/sec]. Without in-plane anisotropy, the maximum velocity should be about 13 m/sec, as can be calculated from the empirical

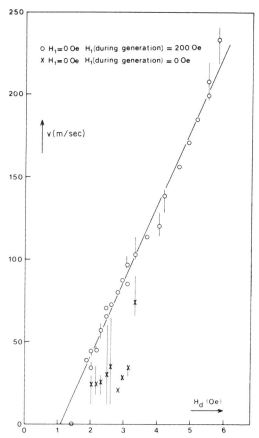

Fig. 23 Bubble translation velocity of film 1. The bubble diameter in this case is 1.8 μm. The drive field H_d is the field gradient × bubble radius. Results are (×) B-type bubbles and (○) A-type bubbles. The mobility in this case is 50 m/sec·Oe, and H_1 is the in-plane field. During the bubble translation measurement, $H_1 = 0$.

equation of de Leeuw (1978). This result clearly shows the influence of the in-plane anisotropy on the maximum velocity. However, such a high Q_2 value could not be realized in the $(Gd, Y)_3(Fe, Mn, Ga)_5O_{12}$ system for device materials with 2-μm and 1-μm bubbles (see Table II). A result for the film with 2-μm bubbles is shown in Fig. 23. In this case we had to face the problem of the occurrence of the A-type and B-type bubbles (see Section 2.3.2.). The velocity measurements were done without an in-plane field. For the B-type bubbles a strong scatter in the bubble velocity was observed, together with the occurrence of a skew angle and skew angle changes. This result is an indication for complicated wall structures.

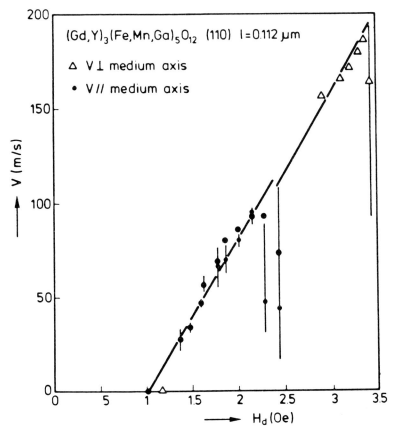

Fig. 24 Bubble translation velocity of film 2. The velocity was measured with the drive field parallel and perpendicular to the medium axis, giving maximum velocities of about 100 and 190 m/sec, respectively.

The A-type bubbles move in the direction of the field gradient, and the velocity is linearly dependent on the drive field as is shown by the curve with the open circles. The maximum velocity of about 200 m/sec is limited by the maximum available drive field.

A similar result for the A-type bubbles was obtained in the film 2 composition as is shown in Fig. 24. In this case a maximum in the velocity was observed. This gave the opportunity to look for the anisotropy in the maximum bubble velocity. The maximum bubble velocity for bubbles moving in the direction of the axis of medium anisotropy energy was 100 m/sec, whereas for bubbles moving in the hard direction the maximum velocity is 190 m/sec. If one assumes that the bubble velocity is determined by the domain-wall segments perpendicular to the translation direction, these

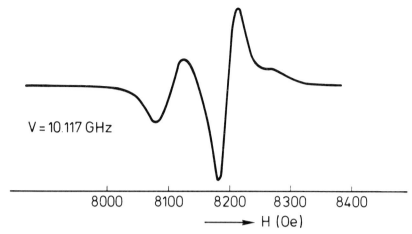

Fig. 25 FMR linewidth of film 2 with the external magnetic field parallel to the in-plane [00$\bar{1}$] direction.

velocities can be compared with v_{max}^b and v_{max}^a [Eqs. (30) and (31)], being 450 m/sec and 690 m/sec, respectively. The discrepancy between theory and experiment is comparable with the discrepancy in the straight domain-wall experiment (Section 2.4.1.). The velocity scatter of the B-type bubbles is quite surprising. In spite of the fact that the drive fields are far below the drive field where a maximum occurs for the A-type bubbles, the velocity and the direction of the translation is not reproducible.

2.4.3 Domain-wall damping

The mobility of the domain wall is given by

$$\mu = (M_s/\lambda')(A/K_u')^{1/2}, \qquad (32)$$

where λ' is the Landau-Lifschitz damping parameter divided by γ^2.

The ferromagnetic linewidth also depends on this damping constant:

$$\Delta H = 2\omega\lambda'/M_s, \qquad (33)$$

where ω is 2π times the ferromagnetic resonance frequency.

In Fig. 25 the resonance line is shown for the film 2 composition. The only indication for a transient layer is the splitting of the resonance line as shown. This shows that the inhomogeneity of K_u/M_s is less than 1%, proving the homogeneity of the (110)-oriented films. In Table 3 the damping constants that can be obtained from FMR at 10^{10} Hz and from the mobility measurements are compared. One sees that the disagreement is less than a

Table 3 Damping constants.

Film[a]	λ' (FMR)	λ' (μ)
Film 1	$11 \times 10^{-9}\,\text{Oe}^2 \cdot \text{sec}$	$18 \times 10^{-9}\,\text{Oe}^2 \cdot \text{sec}$
Film 2	$14 \times 10^{-9}\,\text{Oe}^2 \cdot \text{sec}$	$18 \times 10^{-9}\,\text{Oe}^2 \cdot \text{sec}$
Film 3	$8 \times 10^{-9}\,\text{Oe}^2 \cdot \text{sec}$	$18 \times 10^{-9}\,\text{Oe}^2 \cdot \text{sec}$

[a] Film 1 is the film of Fig. 23, film 2 of Fig. 24, and film 3 of Fig. 22.

factor of two, and in this case FMR linewidth measurements can be used quite well to estimate the bubble mobility.

3 BISMUTH-CONTAINING GARNET FILMS WITH HIGH UNIAXIAL ANISOTROPY AND LOW DAMPING

3.1 Introduction

As shown by Thiele (1971), the optimal device bubble diameter is given by $\phi \simeq 8l$. From Eqs. (1), (2), and (3), it follows that in order to decrease the bubble diameter the value of K_u has to be increased. In known bubble garnet materials, a larger K_u is always brought about by increasing the amount of anisotropic rare-earth ions, which contribute simultaneously to the damping. Due to this contribution, the mobility decreases much faster with decreasing bubble size than predicted by Eq. (32). The use of stress-induced anisotropy is one way of keeping the damping low in films for small bubbles. In fact, in the $(\text{Gd}, \text{Y})_3(\text{Fe}, \text{Mn}, \text{Ga})_5\text{O}_{12}$ films of the preceding section, the damping was nearly independent of the anisotropy. The manganese ions increase λ_{100} much more strongly than λ_{111}. Therefore the optimum value of K_u can be obtained if these films are grown with mismatch on (100)-oriented substrates instead of on (110)-oriented substrates $[K_u = (3/2)\sigma\lambda_{100}]$. It has been shown that in this way anisotropies of $2 \times 10^5\,\text{erg/cm}^3$ can be realized (Breed et al., 1981). More recently, a great deal of attention has been given to a new possibility of introducing anisotropy without the use of damping rare-earth ions, namely, Bi-containing garnet films (Robertson et al., 1981, 1982; Breed et al., 1981; Hibiya, 1981; Hibiya et al., 1981; Luther et al., 1982; LeCraw et al., 1982). Extremely low damping in combination with high K_u values have been realized in these films. Robertson and Voermans of our laboratory and Hibiya et al. (1981) have also grown Bi-containing garnet films with orthorhombic anisotropy.

The obtained values of the in-plane anisotropy, however, appeared to be too small for 1-μm and 2-μm bubble device applications.

3.2 Growth of Bismuth-containing films

The Bi-containing films have been grown by LPE, using the horizontal dipping technique. During growth the films were rotated at about 100 rpm with reversal of rotation after every five revolutions. The films were grown on 2.5-cm-diameter (111)-oriented garnet substrates.

The amount of Bi incorporated into a film increases strongly with decreasing growth temperature. However, the quality of the film also decreases. In order to have growth temperatures that are not too low, a flux system based on Bi_2O_3 as solvent is more favourable than the usual PbO/B_2O_3 flux system (Krumme *et al.*, 1976). The melt composition of the PbO/Bi_2O_3 flux used is given in Table 4. The Bi/Lu and Y/Lu ratios in the

Table 4 Melt composition of Bi-garnet.[a]

Compound	Mole
PbO	1.3
Bi_2O_3	0.26
Y_2O_3	0–0.010
Lu_2O_3	0–0.004
Fe_2O_3	0.17
Ga_2O_3	0–0.023

[a] Depending on melt composition, growth temperatures between 680 and 895 C were used.

films have been used to obtain the desired uniaxial anisotropy and to simultaneously match the lattice parameter of the film to the lattice parameter of the substrate.

Gallium has been used to reduce the saturation magnetization. Different substrates were used in order to increase the film lattice parameter and so to increase the maximum Bi content.

3.3 Anisotropy

The anisotropy of the Bi-containing garnet films increases with the Bi concentration and decreases with magnetic dilution. This is demonstrated in Fig. 26, where the decrease of K_u with Ga concentration is shown for two series of films grown on $Gd_3Ga_5O_{12}$ and $Sm_3Ga_5O_{12}$, respectively. The films were grown without mismatch, so due to the larger lattice parameter more Bi has to be incorporated in the films grown on $Sm_3Ga_5O_{12}$, resulting in a higher value of K_u. The maximum value of K_u that has been obtained

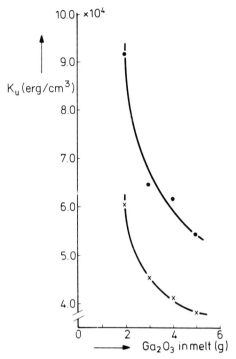

Fig. 26 The anisotropy K_u as a function of the Ga_2O_3 concentration in the melt for films grown on $Gd_3Ga_5O_{12}$ substrates (×) and on $Sm_3Ga_5O_{12}$ substrates (●). The Y_2O_3/Lu_2O_3 weight ratio was 0.9.

at this writing is $1.5 \times 10^5 \, erg/cm^3$ in a film grown on $Nd_3Ga_5O_{12}$ by Robertson et al. (1982). This value of K_u is sufficiently high for submicrometer bubble diameters. The K_u increases by about $8 \times 10^4 \, erg/cm^3$ for one Bi^{3+} ion per formula unit.

3.4 Dynamic properties of the Bismuth-containing garnet films

Since no damping rare earth ions have to be used to increase K_u, the mobility in these films is very high. The maximum velocity in these films with uniaxial anisotropy is only about 20 m/sec, so the drive-field range where the mobility can be measured is too small. In order to be able to measure the mobility, this drive-field range was expanded by increasing the maximum velocity by applying a strong in-plane field. A result for an in-plane field of 1000 Oe is shown in Fig. 27. The film is a $(Bi, Y, Lu)_3(Fe, Ga)_5O_{12}$ film grown on GGG with the parameters $h = 2.76 \, \mu m$, $4\pi M = 311 \, G$, and

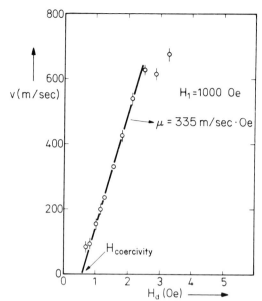

Fig. 27 The velocity of a 3-μm bubble versus drive field in a $(Bi, Y, Lu)_3$-$(Fe, Ga)_5O_{12}$ film with $K_u = 4.4 \times 10^4 \, erg/cm^3$.

$K_u = 4.4 \times 10^4 \, erg/cm^3$. The damping constant obtained from this mobility [see Eq. (32)] is $\lambda' = 1.2 \times 10^{-9} \, Oe^2 \cdot sec$. This damping constant is extremely small and is only twice as large as in pure YIG films. Robertson et al. (1981) showed that the damping obtained from the mobility compares very well with the damping calculated from the FMR linewidth. From the FMR linewidth measurements, it was concluded that mobilities of 600 m/s·Oe can be expected in films with 0.5-μm bubble diameter (Robertson et al. 1982). This is an improvement of a factor 100 over conventional device materials with uniaxial anisotropy.

3.5 Temperature dependence of bubble parameters

Le Craw et al. (1982) improved Bi garnet films of nominal compositions $(Bi, Y, Ca)_3(Fe, Si)_5O_{12}$ for 1-μm bubble diameter devices. It appeared that the temperature dependence of the anisotropy of these films was quite different from the temperature dependence of the conventional (Sm, Lu) garnets. This is shown in Fig. 28.

This different temperature dependence is very favourable and extends the temperature range where bubble devices can be used considerably. A temperature range of -50 to $+150°C$ seems to be possible.

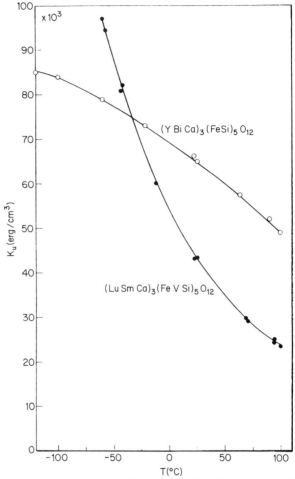

Fig. 28 Temperature dependence of the uniaxial anisotropy of a Bi-containing garnet film compared to a (Sm, Lu) garnet film. (From LeCraw *et al.*, 1982.)

4 DEVICE IMPLICATIONS

4.1 Current access devices on garnet films with orthorhombic anisotropy

At present, the standard operation mode of magnetic bubble devices is the field access mode. In this operation mode, the propagation of the bubbles is accomplished with the aid of a uniform rotating field, which induces sequential field gradients locally in a permalloy overlay pattern or an implantation-defined contiguous-disc structure. The local field gradients provide the moving force acting on the bubbles. The merits of this propagation mode are simplicity and the absence of the need to access each bit separately. All bubbles are moved simultaneously by one rotation field. The limitation of the concept of field access is the necessity to provide a magnetic field of considerable strength (30–60 Oe), rotating at high frequency. Due to excessive dissipation in the field-generating coils, frequencies higher than 300 KHz are hard to realize. It is this problem that actually limits the bit rate of conventional bubble devices and makes them "slow devices", not the intrinsic material properties of the garnet films. In fact, dynamical bubble properties even of conventional garnet compositions are largely sufficient for such low frequencies. Therefore it is clear that there is no need for "faster" materials for field access devices.

This does not hold for current access devices, which have received renewed attention in recent years. The merits of this concept, especially in the promising form of the dual-sheet conductor device (Bobeck *et al.*, 1979), are its ability to handle high frequencies and also its simpler construction due to the absence of field coils. On the other hand power dissipation tends to be high ($\simeq 10\,\mu$W/bit), but improvements have been made recently. In this context, improvements of material properties, especially higher mobilities and lower coercivities, are very meaningful and lead directly to lower power dissipation and higher bit rates. The materials described in the preceding sections have been tailored to meet these demands.

High-frequency bubble propagation experiments have been carried out on an orthorhombic garnet film processed with a dual conductor perforated sheet overlay, using a planar processing technology. To prevent bubble pinning at the edges in the overlay, a 0.2-μm-thick Teflon layer between the bubble layer and the overlay was used. This bubble pinning is caused by overlay stresses that penetrate into the bubble layer. The Teflon layer, which has a very low Young's modulus, is able to relax these stresses to very low values (Verhulst *et al.*, 1981; Breed and Verhulst, 1981). The dual conductor overlay circuit used for the propagation measurements is shown in Fig. 29. The basic propagation period is 8 μm. Bubbles were propagated in the path AB, which is parallel to the current flow. For different frequencies, the bias field limits at different amplitude of the drive current were measured, resulting in the operating margins given in Fig. 30. The area between the upper and the lower curves is the area of successful bubble propagation.

In the case of conventional uniaxial garnets, the maximum propagation frequency is limited to about 2 MHz by the low mobility and low saturation velocity. From Fig. 30, it can be seen that bubbles can be propagated in orthorhombic garnets at frequencies over 10 MHz. The highest propagation rate observed was 20 MHz which corresponds to a bubble velocity of

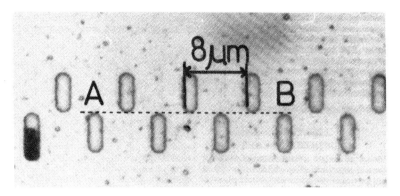

Fig. 29 Top view of a dual conductor perforated sheet overlay for propagation of 1.8-μm bubbles. Owing to the planar technology, the holes in the lower conductor cannot be seen. They are located at the left-hand side next to the holes in the upper conductor. The current flow is parallel to the propagation track AB.

160 m/sec. Higher propagation rates could not be studied due to the limitations of the test equipment, but from Fig. 23 we may conclude that still higher frequencies are possible. The increase of the drive current with frequency is in good agreement with the velocity measurements shown in Fig. 23. From an extrapolation of the curves of Fig. 30, a minimum drive current of 2.4 mA/μm can be predicted for 25 MHz propagation frequency.

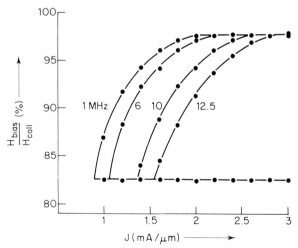

Fig. 30 Operating margins of the dual-conductor circuit of Fig. 29 on a garnet with orthorhombic anisotropy. The bias field on the vertical axis is given relative to the free bubble collapse field. The drive current J on the horizontal axis is equal to the sheet current divided by the width of the sheet.

In conclusion, we may state that a fast bubble device is feasible by combining the current access technique with orthorhombic garnet films of high mobility and high maximum velocity.

4.2 Device potential of high-mobility Bismuth-substituted garnets for bubble diameters <1 µm

The potential usefulness of uniaxial Bi-substituted garnets for bubbles of diameter <1 µm was demonstrated by Luther *et al.* (1982). The test circuit was a 4-µm period contiguous disk pattern using a 1-µm bubble film. The film was implanted with a triple dose, Ne–Ne–H_2. At room temperature the bias field margin was 39 Oe at 25 Oe quasistatic drive field and 50 Oe at 40 Oe drive. Due to the superior temperature behaviour of these materials at higher and lower temperature, an extended temperature operation range is feasible. The domain visibility was the best ever observed for 1-µm bubbles.

Our conclusion is that Bi-substituted garnet films are very suitable materials for high-density bubble devices, both of the contiguous disk and of the permalloy overlay type. In current access devices these materials allow for low power dissipation.

5 CONCLUSION

Mn^{3+}-containing garnet films of high quality can be grown on slightly misoriented (110) GGG substrates. These films have excellent dynamic properties (high maximum bubble velocity and high bubble mobility) for device application. The possibility of high-frequency device operation has been demonstrated. The flexibility in this material system is sufficient to realize a good temperature stability of the static bubble properties. We have shown that the magnetic properties in this system can be predicted very well from the film composition, allowing for accurate materials tailoring.

The Bi-containing films with uniaxial anisotropy have an extremely low damping; also, the anisotropy is sufficiently large for submicrometer bubbles. From a practical point of view, the most interesting property of these films may be the extended useful temperature range (LeCraw *et al.*, 1982).

ACKNOWLEDGEMENTS

The authors would like to emphasize that the work presented in this paper is a result of the work of many colleagues in the Philips Research Laboratories.

REFERENCES

Bobeck, A. H., Blank, S. L., Butherus, A. D., Ciak, F. J., and Strauss, W. (1979). *Bell Syst. Tech. J.* **58**, 1459.
Breed, D. J. and de Geus, W. (1983). *J. Appl. Phys.* **54**, 5314.
Breed, D. J., and Verhulst, A. G. H. (1981). *Microelectron. J.* **12**(5), 15.
Breed, D. J., and Voermans, A. B. (1980). *IEEE Trans. Magn.* **MAG-16**, 1041.
Breed, D. J., Stacy, W. T., Voermans, A. B., Logmans, H., and van der Heijden, A. M. J. (1977). *IEEE Trans. Magn.* **MAG-12**, 1087.
Breed, D. J., van der Heijden, A. M. J., Logmans, H., and Voermans, A. B. (1978). *J. Appl. Phys.* **49**, 939.
Breed, D. J., de Leeuw, F. H., Stacy, W. T., and Voermans, A. B. (1978–1979). *Philips Tech. Rev.* **38**, 211.
Breed, D. J. de Geus, W., and Enz, U. (1980). *J. Appl. Phys.* **50**, 2780.
Breed, D. J., Robertson, J. M., Algra, H. A., van Bakel, B. A. H., de Geus, W., and Heijnen, J. R. H. (1981). *Appl. Phys.* **24**, 163.
Breed, D. J., de Geus, W., Voermans, A. B., and van Bakel, B. A. H. (1982). *J. Appl. Phys.* **53**, 2546.
Breed, D. J., Voermans, A. B., Nederpel, P. Q. J. and van Bakel, B. A. H. (1983). *J. Appl. Phys.* **54**, 1519.
Breed, D. J., Nederpel, P. Q. J. and de Geus, W. (1983). *J. Appl. Phys.* **54**, 6577.
Burton, W. K., Cabrera, N., and Frank, F. C. (1951). *Philos. Trans. R. Soc. London, Ser. A* **243**, 299.
Callen, H., and Josephs, R. M. (1971). *J. Appl. Phys.* **42**, 1977.
de Leeuw, F. H. (1973). *IEEE Trans. Magn.* **MAG-9**, 614.
de Leeuw, F. H. (1977). *IEEE Trans. Magn.* **MAG-13**, 1172.
de Leeuw, F. H. (1978). *IEEE Trans. Magn.* **MAG-14**, 596.
de Leeuw, F. H., van den Doel, R., and Enz, U. (1980). *Rep. Prog. Phys.* **43**, 689.
Dionne, G. F. (1979). *J. Appl. Phys.* **50**, 8257.
Dionne, G. F., and Goodenough, J. F. (1972). *Mater. Res. Bull.* **7**, 749.
Dorleijn, J. W. F., Druyvesteijn, W. F., Bartels, G., and Tolksdorf, W. (1973). *Philips Res. Rep.* **28**, 133.
Eschenfelder, A. H. (1977). *J. Appl. Phys.* **49**, 1891.
Eschenfelder, A. H. (1980). "Magnetic Bubble Technology", Springer-Verlag, Berlin and New York.
Gerhardstein, A. C., Wigen, P. E., and Blank, S. L. (1978). *Phys. Rev. B* **18**, 2218.
Hagedorn, F. B. (1970). *J. Appl. Phys.* **41**, 1161.
Hagedorn, F. B. (1972). *AIP Conf. Proc.* No. 5, 72.
Hansen, P. (1974). *J. Appl. Phys.* **45**, 3638.
Hibiya, T. (1981). *J. Appl. Phys.* **52**, 4720.
Hibiya, T., Makino, H., and Konishi, S. (1981). *J. Appl. Phys.* **52**, 7347.
Konishi, S., Engemann, J., and Heidmann, J. (1981). *Appl. Phys. Lett.* **36**, 467.
Krumme, J. P., Bartels, G., Hansen, P., and Robertson, J. M. (1976). *Mater. Res. Bull.* **11**, 337.
LeCraw, R. C., Luther, L. C., and Gyorgy, E. M. (1982). *J. Appl. Phys.* **53**, 2481.
Luther, L. C., LeCraw, R. C., Dillon, J. F., and Wolfe, R. (1982). *J. Appl. Phys.* **53**, 2478.
Malozemoff, A. P., and Slonczewski, J. C. (1979). "Magnetic Domain Walls in Bubble Materials", Applied Solid State Science, Suppl. 1. Academic Press, New York.

Pierce, R. D. (1972). *AIP Conf. Proc.* No. 5, 81.
Robertson, J. M., Algra, H. A., and Breed, D. J. (1981). *J. Appl. Phys.* **52**, 2338.
Robertson, J. M., Breed, D. J., and Algra, H. A. (1982). *J. Appl. Phys.* **53**, 2483.
Röschmann, P. (1980). *J. Phys. Chem. Solids* **41**, 569.
Röschmann, P., and Hansen, P. (1981). *J. Appl. Phys.* **52**, 6257.
Rossol, F. C. (1971). J. de Phys. **32**, C1, 437.
Schlömann, E. (1976). *J. Appl. Phys.* **47**, 1142.
Slonczewski, J. C., Malozemoff, A. P., and Giess, E. A. (1974). *Appl. Phys. Lett.* **24**, 396.
Stacy, W. T., Voermans, A. B., and Logmans, H. (1976). *Appl. Phys. Lett.* **29**, 817.
Strocka, B., Holst, P., and Tolksdorf, W. (1978). *Philips J. Res.* **33**, 186.
Thiele, A. A. (1971). *Bell Syst. Tech. J.* **50**, 727.
Van Erk, W. (1979). *J. Cryst. Growth* **46**, 539.
Van Erk, W., van Hoek-Martens, H. J. G. J., and Bartels, G. (1980). *J. Cryst. Growth* **48**, 621.
Vella-Coleiro, G. P., and Tabor, W. J. (1972). *Appl. Phys. Lett.* **21**, 7.
Verhulst, A. G. H., Bril, T. W., Voermans, A. B., and Koel, G. J. (1981). *J. Appl. Phys.* **52**, 2371.
Voermans, A. B., Breed, D. J., van Erk, W., and Carpay, F. M. A. (1979). *J. Appl. Phys.* **50**, 7827.
White, R. L. (1973). *IEEE Trans. Magn.* **MAG-9**, 606.
Yatsenko, V. A., Bokov, V. A., Bystrov, M. V., Sher. E. S., and Trofimova, T. K. (1979). *Sov. Phys. Solid State* (*Engl. Transl.*) **21**, 1528.

4 Field Access Permalloy Devices

SHOBU ORIHARA and TAKEYASU YANASE

Fujitsu Laboratories
Kawasaki, Japan

1	Introduction	138
2	Problems in higher density devices	139
	2.1 Increase in drive field	140
	2.2 Other problems	142
3	Design of function elements	143
	3.1 Stretcher	144
	3.2 Replicator/replicate gates	148
	3.3 Transfer gates/swap gates	153
	3.4 Generator	157
	3.5 Detector	161
4	Optimization of propagation	165
	4.1 Bubble diameter and pattern design	165
	4.2 Pattern period, gap, and spacing	167
	4.3 Effect of crystallographic orientation	171
	4.4 Conductor crossing	174
5	Loop organization	178
6	Design and characteristics of a 1-Mbit chip	184
	6.1 1-Mbit chip design	184
	6.2 1-Mbit chip characteristics	186
7	Further development of permalloy devices	193
	7.1 New technologies for higher density devices	193
	7.2 8-μm period "wide-gap" track and its scaling	196
	7.3 Optimization of 4-μm period "wide-gap" track	200
	7.4 Design and characteristics of a 4-μm period major–minor chip	207
	References	210

1 INTRODUCTION

The first practical bubble memory device was realized by employing permalloy bubble propagation patterns (Bonyhard *et al.*, 1973). A 20-kbit memory chip was designed with 28-μm period T-bar propagation patterns, a nucleation generator, chevron stretchers, a thick-film serpentine detector, dollar-sign transfer gates, and a major–minor loop organization. The chip was fabricated using 6 μm diameter bubble LPE film with a permalloy-first process. Following this design, several memory devices were developed to be evaluated for practical applications; these were 10–20 kbits in capacity, 20–28 μm in propagation period, 5–6 μm in bubble diameter, and had single loop and major–minor loop organizations. All of these devices were designed based on the T-bar pattern design rule, which is reviewed by Bobeck and Della Torre (1975). Although chips with the higher capacities of 64 kbits (Hiroshima *et al.*, 1976) and 92 kbits (Naden *et al.*, 1976) were designed using the T-bar pattern with a propagation period of about 20 μm in larger chip formats, some problems were encountered with the T-bar pattern when the propagation period was reduced to 16 μm to realize chip capacities of 64 kbits or more without increasing the chip size. In the T-bar pattern, pattern gap, which is the minimum feature size, should be held to less than one-sixteenth of the propagation period. With a 16-μm period, pattern gap is around 1 μm; this is difficult to achieve with conventional photolithography. Even with a 1-μm gap, 16-μm period T-bar propagation patterns require a very high drive field to propagate 3-μm diameter bubbles because of self-stabilizing coupling between permalloy patterns and bubbles with higher magnetization (Igarashi *et al.*, 1976).

These problems were completely solved by the epoch-marking gap-tolerant pattern designs of Bonyhard and Smith (1976) and Bobeck and Danylchuk (1977). Their designs are shown in Fig. 1. These are called the half-disk pattern and asymmetric chevron pattern, respectively. In these gap-tolerant patterns, the minimum feature size can be increased to twice that of the T-bar pattern; this is about one-eighth of the propagation period, and the drive field required for 16-μm period propagation with 3-μm bubbles is as low as that for 28-μm T-bar propagation with 6-μm bubbles. Moreover, important functions such as block-replicate gates and swap gates can be easily realized with the half-disk pattern (Bonyhard and Smith, 1976, Bonyhard, 1977; Gergis and Kobayashi, 1978); this is to be expected, since the half-disk pattern originated from a pickax pattern, which works very well as a replicator and generator (Bonyhard *et al.*, 1974). Making use of the possibilities of these gap-tolerant patterns, commercial memory chips with capacities of 64–256 kbits were developed with various loop organizations: i.e., serial loop (Bobeck, 1977), major–minor loop (Orihara *et al.*, 1978), and block-replicate (Gergis *et al.*, 1979b; Bullock *et al.*, 1979b).

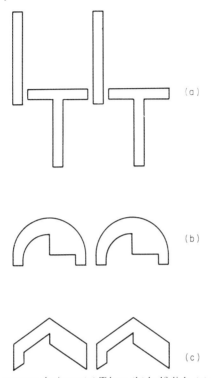

Fig. 1 Propagation pattern designs: (a) T bar, (b) half disk, (c) asymmetric chevron.

The pursuit of even higher bit density is inevitable in bubble memory technology, as well as in other memory technologies. The possibility of realizing 1-Mbit chips was foreseen when the gap-tolerant pattern design appeared, and this was demonstrated as early as in 1977 (Archer, 1977). Practical 1-Mbit chip designs, however, require further development of design and process technologies. This part describes high-density permalloy device designs, problems encountered when the propagation period is reduced to 8 μm or less, design considerations for solving these problems, the design and characteristics of a 1-Mbit chip, and prospects of realizing higher density permalloy devices.

2 PROBLEMS IN HIGHER DENSITY DEVICES

This section discusses problems encountered in designing higher density permalloy devices using the example of an 8-μm period 1-Mbit device designed by scaling down from a 16-μm period 256-kbit device, whose design is established well enough to be applied for practical purposes.

2.1 Increase in drive field

It is known that the minimum drive field required to propagate bubbles tends to increase as the bubbles are made smaller. This is due to the increased polarizing effect of the bubbles. As the bubble diameter is reduced, magnetization of the bubbles increases proportionally; the increased magnetic fields of the bubbles polarize permalloy patterns, so that a stronger drive field is required to overcome the polarizing effect for bubble propagation. Linear dependence of the minimum drive field on the saturation magnetization of the bubbles has been predicted by modeling (George et al., 1974) and confirmed experimentally on T-bar propagation patterns (Kryder et al., 1974). Although the superior propagation characteristics of gap-tolerant patterns appear to be due to the difference in this polarizing effect— the gap-tolerant patterns are more resistant to the polarization than the T-bar pattern—the minimum drive field required for the gap-tolerant pattern also increases as the saturation magnetization is increased. Figure 2 shows the dependence of the drive field in an 8-μm period half-disk propagation track on saturation magnetization of 1.5-μm diameter bubble films (Inoue et al., 1979).

Besides the increase in drive field required for bubble propagation, other factors which increase the drive field required in overall memory operation should be considered. For example, Fig. 3 shows operating margin curves in

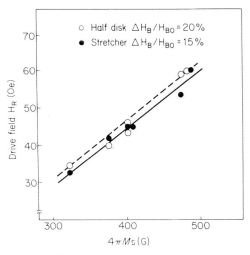

Fig. 2 The $4\pi M_s$ dependence of the drive field which is necessary to produce normalized bias margin of 20% for half-disk propagation tracks or 15% for chevron stretchers (1.5 μm bubbles, 8-μm propagation period, 100-kHz sinusoidal drive).

an 8-μm period memory chip (Inoue, 1979), which was designed by scaling down from a 16-μm period device including the following functions: 8-μm period half-disk propagation elements with a 1-μm minimum feature size, a pickax generator, block replicate gates, chevron stretchers, a serpentine detector, etc. (Bonyhard and Smith, 1976). As shown in Fig. 3, the minimum drive field required for overall operation is determined not by propagation but by stretching and detector noise. In the 16-μm period devices, design of these functions was optimized to avoid significantly limiting overall operating margins, but greater optimization seems to be needed for 8-μm period devices. Also, bubble stretching (which is required for bubble replication as well as for detection) and occurrence of detector noise should be investigated, since these are quite different matters from bubble propagation.

Another important factor which increases the drive field required is bubble propagation across conductor steps. Bubbles are trapped at conductor edges under permalloy patterns because of unwanted magnetic poles at the

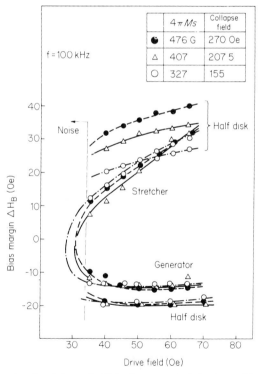

Fig. 3 Bias margins for 1.5-μm bubble, 8-μm period devices with different $4\pi Ms$ (100–kHz sinusoidal drive). Below 35 Oe, margins cannot be measured due to large switching noise from detectors.

permalloy pattern steps. This conductor step problem appears to become more serious when the bubble diameter and propagation period are reduced, and any wafer processing measures for planarizing the conductor steps should be employed in high-density devices.

The required amplitude of rotating drive field differs according to the coil drive current waveform. The circular locus of a field rotated by sinusoidal current is an original and ideal drive method for permalloy devices, but the diamond-shape locus produced by triangular current is more popular in commercial bubble memory devices because of the simplicity of the coil drive circuits. The nominal drive fields used in the 16-μm period devices were 40–50 Oe for sinusoidal drive current and 50–60 Oe (peak field) for the triangular drive current. The 8-μm devices are expected to require drive fields which are 25–30% larger. The increase in drive field not only increases power consumption of the device but limits drive frequency and operation temperature range. All the data in the following chapters are based on the triangular drive waveform and are expressed in terms of peak field.

2.2 Other problems

The minimum feature size of the 8-μm period propagation pattern should be 1 μm or less when it is scaled down from a 16-μm period pattern with a 2-μm minimum feature size. Fine pattern delineation is easier in bubble memory devices than in semiconductor devices because wafer processing is much simpler and dry etching can be easily adopted.

The delineation of 1-μm patterns in 8-μm period devices, however, is rather difficult to obtain with conventional photolithography (mainly because of its low resolution) and requires the development of new technologies. Variations in pattern gap size result in a considerable increase in drive field required (Bonyhard, 1979). Of the new photolithographic technologies, projection lithography could be satisfactory for 8-μm bubble device processing (Matsuyama et al., 1977), and this method is widely used (Bullock et al., 1979b; George et al., 1981a; Yanase et al., 1981). However, it is hoped that new pattern designs with less stringent minimum feature size requirements will be developed to allow higher density devices with propagation periods smaller than 8 μm to be realized.

Another problem with lithography is registration between conductor patterns and permalloy patterns. From experience with 16-μm period devices, registration tolerance might be less than $\pm 0.25\,\mu$m for scaled 8-μm period devices. This value does not seem easy even for the new lithographic technologies, so the function patterns should be designed with sufficient registration tolerances.

Current densities in the conductor patterns of bubble memory devices are fairly high (on the order of $10^7 \, A/m^2$) and the electromigration problem should be taken into consideration in order to guarantee reliability. In 16-μm period devices, conductors are usually AlCu with a minimum width of about $4\,\mu$m. The reliability of such devices has been confirmed to be satisfactory (Yamaguchi and Hibi, 1980). In the scaled 8-μm period device, however, the reliability of conductors with a minimum width of 2-μm has yet to be studied, and new materials such as Au or Cu have been discussed for replacing AlCu. Conductor width should be kept as large as possible in the function pattern design so as to reduce the current density.

Besides the problems caused by reduced period or minimum feature size, serious problems are encountered when the capacity is increased (for example) from 256 kbit to 1 Mbit. An inherent problem is the deterioration in access time. As a bubble memory is basically a kind of shift register, the access time increases in proportion to the square root of the capacity unless the drive frequency is increased. For example, average access time increases from 6 msec with 256-kbit devices to 12 msec with 1-Mbit devices. Another problem to be considered is the increase in function drive voltages. The voltage to drive replicate gates is fairly high and increases with the number of gates. Sometimes gates are divided and driven in parallel with limited voltages. The replicate-gate conductor should be as wide as possible in order to reduce its resistance, as well as to improve its reliability.

The yield of defect-free chips is generally lower in higher density devices. In bubble devices, however, defect-tolerant schemes have been employed; in multi-loop devices, defective minor loops are replaced by additional redundant loops under the management of control electronics in memory systems. In megabit devices, the yield problem can be settled by this scheme. Finally, the stability of bubbles at stop positions appears to be poorer in higher density devices, although it has not yet been determined whether this problem is an inherent one. When drive-field rotation is stopped and the drive field is removed, the stability of stopped bubbles is ensured by the application of a small dc in-plane field, which is usually provided by a slightly inclined vertical bias field. With smaller propagation periods, larger in-plane holding fields tend to be required, resulting in an increase in drive field required.

3 DESIGN OF FUNCTION ELEMENTS

The design of function elements (stretcher, replicate gates, swap gates, generator, and detector) is discussed in this section. Some examples of former designs and design considerations are described, and improved designs for 8-μm period 1-Mbit devices are presented.

3.1 Stretcher

The stretcher is a function for stretching bubbles to long stripe domains to increase the magnetic flux of bubbles applied to the detector. The multi-bar chevron pattern allows unique propagation of stripe domains elongated laterally to propagation tracks (Bobeck et al., 1972). Scores of columns, each of which consists of hundreds of chevrons, are used for thick film permalloy detectors (Bobeck et al., 1973). The number of chevrons is chosen to obtain the required output voltage in the detector, while the number of columns is determined to provide propagation steps (time) sufficient for the bubble to be fully stretched. An original chevron stretcher design is shown in Fig. 4a, together with some modified designs.

As shown in Fig. 3, the upper threshold of the bias field margin in bubble stretching is generally lower with low drive fields and depends more strongly on the drive field than that in propagation. This is due to the following differences in the propagation and stretching mechanisms. Bubbles are transported by the local gradients of potential wells under the permalloy pattern, and these gradients are not strongly dependent on the drive field. Bubbles are stretched, however, when the absoluted depth of the potential wells exceeds the bubble strip-out threshold; this potential well depth requires a drive field which is larger in almost linear proportion. In Fig. 3 the bias margin of propagation increases in proportion to the saturation magnetization of the bubble film. This is due to the fact that the bias field which provides the minimum bubble diameter for propagation at a fixed drive field is considered to be proportional to the free bubble collapse field, and the bubble strip-out field along propagation tracks is also proportional to the free bubble strip-out field. The free bubble stability range between the collapse and strip-out fields has a linear relationship to the saturation magnetization of the film. On the other hand, the bias margin curves of the stretcher are almost the same for the different magnetization films; threshold is determined solely by the potential well depth and the free bubble strip-out field, because the potential depth from the stretcher permalloy patterns to stretch bubbles is independent of the collapse field.

The dependence of bias margin on saturation magnetization is considered applicable when the saturation magnetization increases along with the reduction in bubble diameter. Thus, the bias margin for bubble stretching tends to be even smaller with low drive fields than for propagation when the bubble diameter is reduced to obtain higher densities. In the design of 8-μm period devices, it is essential to obtain good stretch margins at low drive fields; in other words, by improving the stretch margin, the drive field for overall operation of 8-μm period devices is expected to be held to the same level as with 16-μm period devices.

Field Access Permalloy Devices 145

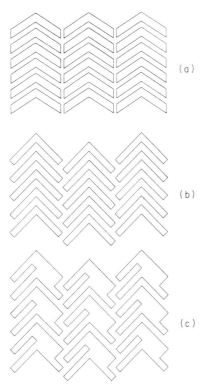

Fig. 4 Chevron stretcher designs.

The modified chevron stretchers shown in Fig. 4b and 4c were devised to improve the stretch margins at low drive fields. The drive field required to obtain a certain bias margin in the design in Fig. 4c is about 5 Oe lower than in Fig. 4b, which has an 18-μm column period for stretching 3-μm diameter bubbles (Yanase *et al.*, 1978). The column period dependence of the stretch margin curves in Fig. 4c is shown in Fig. 5a and b for different process conditions (Yanase *et al.*, 1979); spacing between the garnet film and permalloy patterns is 6600 Å in Fig. 5a and 4500 Å in Fig. 5b, and permalloy thickness is 4500 Å in Fig. 5a and 3700 Å in Fig. 5b. The column period is varied from 10.5 μm to 15.0 μm for stretching 1.5-μm diameter bubbles, and the data is compared with the margin curve for propagation in the 8-μm asymmetric chevron track shown in Fig. 23. The minimum pattern width and the gap is 1 μm in all cases. The data is for a 100-kHz triangular drive, and the upper thresholds correspond to the bias fields with which output voltage from the detectors is half that of fully stretched

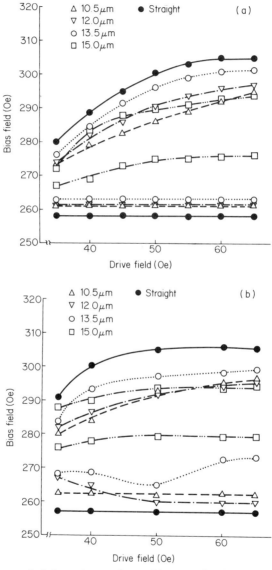

Fig. 5 Column period dependence of stretching margins in the design of Fig. 4c, compared with propagation margin in an 8-μm period asymmetric chevron track (1.5-μm bubbles, 100-kHz triangular drive): (a) 6600 Å spacing and 4500 Å permalloy; (b) 4500 Å spacing and 3700 Å permalloy.

bubbles. The upper threshold of the stretch margin tends to be improved initially by increasing the column period in both Fig. 5a and Fig. 5b. This is expected since the magnetic poles at the chevron ends, which are the main force stretching bubbles, become stronger as the shape anisotropy of the chevron pattern increases according to the increase in column period. In 16-μm period devices, slightly longer column periods have been used, for example, 20 μm (Bobeck and Danylchuk, 1977). However, both the upper and lower threshold of the bias margin are degraded for a much longer period of 15 μm, although slight improvements are observed for the upper threshold at low drive fields (as shown in Fig. 5a and 5b). This deterioration is attributed to propagation failure as stretched domains traverse the long arms of the chevron patterns. For the structure with thicker layers (Fig. 5a), this degradation is not observed for the periods up to 13.5 μm. However, for the structure with thinner layers, the lower threshold of the stretching margin at high drive fields begins to deteriorate at a period of 13.5 μm. This is attributed to the difference in permalloy film thickness rather than to the difference in spacing; the weaker shape anisotropy of the thicker permalloy provides stronger magnetic poles along the long arms of the chevron patterns, so stripe domains are propagated more smoothly at the long period. On the other hand, the spacer thickness can explain the difference in upper thresholds at low drive fields in Fig. 5a and 5b. In Fig. 5a, the propagation margins of the 8-μm asymmetric chevron track, as well as the stretch margins, are narrow at low drive fields. Considering the rather moderate dependence of propagation on spacing, the 6600 Å spacing seems to be a bit too large for 1.5-μm diameter bubbles or 8-μm period propagation. Smaller spacing and thicker permalloy film appear preferable as far as the stretcher is concerned; however, these factors are determined by the detector design and by step coverage processing as described later.

The data in Fig. 5 led to the new stretcher design shown in Fig. 6, which is called the Mt. Fuji stretcher. This design is intended to improve stripe propagation along the long-period chevron arms. Figure 6 also shows the stretch margins of the Mt. Fuji stretcher with a column period of 15 μm for two different processing structures, together with the bias margins for 8-μm straight propagation. Not only is the lower threshold deterioration in stretch margin eliminated with a period of 15 μm, but the upper thresholds are nearly the same as for straight propagation. The difference in processing is attributed to the difference in spacing. With a spacing of 4500 Å, a sufficient bias margin (40 Oe) is obtained with a 55-Oe 100-kHz triangular drive field.

In this stretcher design, another type of error is sometimes found at low bias fields, depending on the garnet film parameters, spacing, and so forth. This type of error is caused by the phase lag of propagating stripe ends at the bottom of the stretcher columns, and can be effectively avoided by

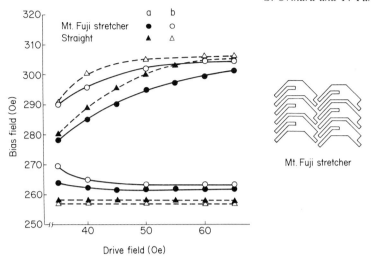

Fig. 6 Stretcher margins of 15-μm period Mt. Fuji stretcher compared with 8-μm period straight propagation margin for different structures same as in Fig. 5 (1.5-μm bubbles, 100-kHz triangular drive).

adding some original chevron patterns to the bottom of the Mt. Fuji stretcher columns.

Although various design variations are possible (Bullock et al., 1979a; George et al., 1981a), the stretcher design with a column period which is twice as long as the propagation period in minor loops is considered to be a powerful design concept for improving bias margins and reducing the stretcher drive fields in 8-μm period devices. Further, this long-period stretcher is expected to be useful in devices of even higher density.

3.2 Replicator/replicate gates

Replication is the process of splitting a bubble in two for copying stored data prior to detection, and is required for non-destructive read-out in memory devices with destructive read-out type detectors (such as thick-film serpentine detectors. Block replication (simultaneous replication of bubbles read out from each minor loop) has been recently employed in one standard chip architecture, mainly because it facilitates control of read and write operations.

An original replicator design with excellent performance is the pickax replicator shown in Fig. 7a (Bonyhard et al., 1974; Bonyhard and Smith, 1976). A bubble propagated to the head of a pickax (A) is stretched across

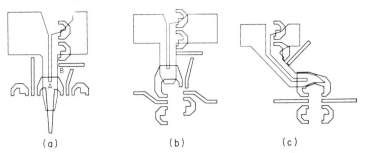

Fig. 7 Replicator and replicate gate designs: (a) pickax, (b) $\frac{5}{4}\pi$, (c) sideways.

the top by a strong and widely spread magnetic pole which is provided by the long leg of the pickax. A strong current pulse (cut pulse) applied through a hairpin conductor cuts the stretched bubble by generating a vertical field which is increasing bias field inside the conductor loop. A subsequent pulse (transfer pulse) with the same polarity and a smaller amplitude holds the trailing bubble in position until an attractive pole is formed at the end of bar pattern B. Then the bubble is transferred and starts to propagate along the new propagation path, while the leading bubble continues to propagate along the original path. Although this pickax replicator provides excellent performance, good bias margin, and wide pulse amplitude and phase margins, the strong magnetic pole which is formed at the end of the long leg at the opposite rotating field phase affects propagation of neighboring bubbles and degrades the bias margin when a replicator is located at the end of each closely packed minor loop for block replication (Bonyhard and Smith, 1976). Some modified replicate gate designs, shown in Fig. 7b and 7c, have been developed to overcome this problem (Gergis and Kobayashi, 1978; and Bonyhard et al., 1977). In these designs, strong shape anisotropy is attained by elongating the pattern along the bubble propagation path.

For stable replication, a bubble should first be stretched long enough to be cut by a succeeding current field, and conditions should be the same as in the stretcher when the bubble diameter is reduced. In order to obtain good bias margins at low drive fields, it is necessary to provide a stronger magnetic pole across the hairpin conductor pattern. From the successful result obtained in the long period stretcher, improvement in the replicate gate design can also be expected by utilizing a permalloy pattern which is larger than that scaled from the design shown in Fig. 7 (Orihara et al., 1979). Improvement of the replicate gate for 1.5-μm diameter bubbles by enlarging the pickax pattern is shown in Fig. 8a. The width of the pickax is increased from 9 to 15 μm. The 9-μm width is half the width of the pickax in the 3-μm bubble device. In Fig. 8b, the bias margins of these replicate gates are shown against replicate pulse phase at a drive field of 45 Oe. Because of the

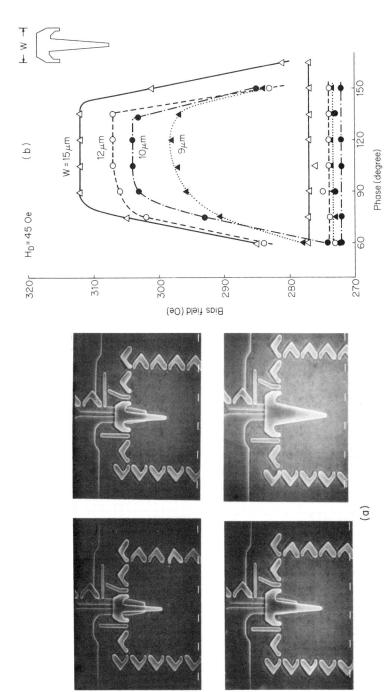

Fig. 8 (a) Experimental replicate gate designs with 9–15 μm pickax width, and (b) bias margins versus replicate pulse phase (1.5-μm bubbles; 45-Oe, 100-kHz triangular drive field).

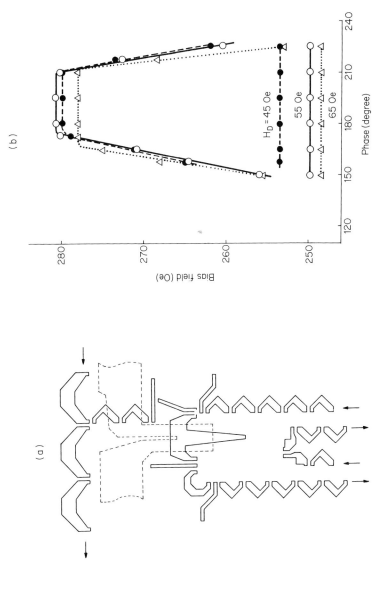

Fig. 9 (a) Replicate gate design with a 13.5-μm pickax and a double period major line for an 8-μm period device and (b) bias margin versus replicate phase at 200-kHz triangular drive field of 45, 55, and 65 Oe.

stronger and more widely spread magnetic pole induced along the pickax head by the larger pattern size, significant improvement of the bias margin is obtained at this low drive field, together with some improvement in the phase margin. The margin loss with a low bias field in the 15-μm pickax is due to stripe-out of bubbles at the left arm to the side of the leg which is too wide.

The results obtained in the enlarged replicate gate and long-period stretcher indicate the possibility of reducing the drive field required for 8-μm period devices to a level equivalent to that for 16-μm period devices, as long as spatial restrictions on the functions are relaxed and their pattern periods are increased to approximately twice that of the storage loops, which are maintained at 8-μm. An example of a replicate gate design with an enlarged pickax, double period gates, and a double period major line is shown in Fig. 9, together with its characteristics. The drawback of the long period and enlarged pattern design might be a lower frequency limit.

Thus far, drive frequencies used in practical applications have ranged from 50 to 200 kHz; this is determined by considerations of drive electronics and power dissipation and is not limited by bubble propagation speed. Bubble mobilities of 300–500 cm/sec·Oe in YSmLuCaGeIG films (the

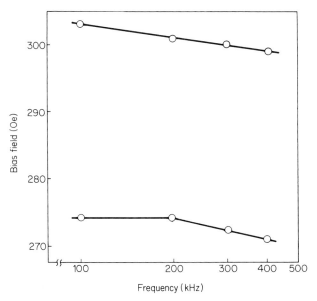

Fig. 10 Frequency dependence of bias margin in a long-period propagation path with 16-μm period asymmetric chevrons and 13.5-μm period Mt. Fuji stretchers. The stripe width and the mobility of the bubble film are 1.5 μm and 360 cm/sec·Oe, respectively.

standard composition of 1.5-μm to 3-μm diameter bubble materials) possess frequency limits which are much higher than these frequencies. Figure 10 shows the frequency dependence of the bias field margin for a 16-μm period propagation path and 13.5-μm Mt. Fuji stretcher. The stripe width and mobility of the LPE film are 1.5 μm and 360 cm/sec·Oe, respectively. Margin deterioration does not occur up to 400 kHz for a 45-Oe drive field. The shift of the thresholds to the low bias field as the drive frequency increases is due to the higher chip temperature resulting from the increase in drive coil power dissipation.

On the other hand, the enlarged gate pattern design allows the use of wide conductor patterns and reduces problems with high gate drive voltages and electromigration. Further, registration tolerance between the conductor pattern and permalloy pattern is expected to be greater.

3.3 Transfer gates/swap gates

Three types of bubble switches have been used for switching bubbles between minor loops and a major loop or major lines. Examples of these switch designs, based on the gap-tolerant half-disk pattern, are shown in Fig. 11. Figure 11a shows a transfer gate which is used for storing newly written bubbles into minor loops from a write major line in a block-replicate–transfer chip configuration (Bonyhard and Smith, 1976). A bubble propagated along the major line is blocked before it crosses hairpin conductor A by a vertical field, which is generated by a current pulse with a polarity which collapses the bubble inside the hairpin conductor. The bubble is held there until an attractive pole is formed at the end of bar pattern B; then it jumps from A to B before the current terminates. Then the bubble begins to propagate

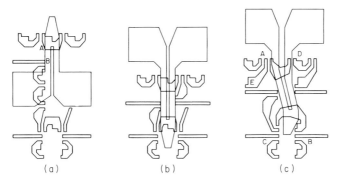

Fig. 11 Transfer-gate and swap-gate designs: (a) transfer switch, (b) bi-directional transfer gate, (c) true swap gate.

along a new path which merges into a minor loop. Figure 11b shows a bi-directional transfer gate (Bonyhard and Smith, 1976) which is used for switching bubbles between a major loop and minor loops for read–write operation in a major–minor loop chip configuration. This gate consists of two combined sets of the transfer gate shown in Fig. 11a, and its transfer-in and transfer-out operations (switching bubbles from the major loop to the minor loops, and vice versa) are the same as those of the uni-directional gate, except that the transfer-out operation is 180 degrees out of phase with the transfer-in operation. This gate can be operated as a swap gate (Bonyhard and Smith, 1976) by exchanging a bubble in the major loop for one in the minor loop in one operation. For this operation, the hairpin conductor current must last long enough for both the transfer-out and transfer-in operations to be completed. The swap gate is useful in reducing read–write cycle time and to facilitate control of write operations. This is because swap gates eliminate the need to carry out erase operations prior to write operations. On the other hand, the swap gate shown in Fig. 11c is sometimes called a true swap gate (Bonyhard, 1977). While the swap gate shown in Fig. 11b does not place the bubble taken out of the minor loop into the position vacated by the in-going bubble, swap operation of gate of Fig. 11c is true in that not only does the in-going bubble go into the position vacated by the out-coming bubble, but vice versa. As shown in Fig. 11c, bubbles at A and B are exchanged and go to C and D, respectively, after two cycles of field rotation.

The most important point in transfer-gate or swap-gate design is to obtain a good bias field margin against gate current amplitude. The timing of the current pulse is not critical as long as its leading and trailing edges cover a core of time during which the bubble is blocked and switched to the bar pattern between propagation paths. Figure 12a shows a bi-directional transfer gate design based on the original design (Bonyhard and Smith, 1976). Figure 12b shows bias field margins against pulse current amplitude for transfer-in and transfer-out operations of this gate in a 16-μm period device. For both operations, current margins are very poor and the low bias field margins are largely lost at high currents. Types of errors which occur in the high current–low bias region are bubble trapping at the entrance of gate half-disk patterns A and B and bubble stripe-out to the end of canted bars C and D. These errors are caused mainly by the polarizing effect induced in the gate half disk patterns by the inplane field generated by the current through the conductor. A bubble approaching the conductor edge is pulled back to point A or B by the attractive pole which is generated by the current field (not by the rotating field), and sometimes the bubble stripes out to bar C or D. By calculation, the strength of this inplane field is more than 10 Oe/10 mA just above the conductor; the cause is confirmed by the

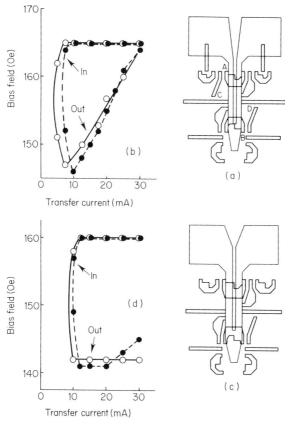

Fig. 12 Transfer-gate designs and current margins. (a) Early design for 16-μm period device and (b) bias margins versus current amplitude. (c) Improved design and (d) characteristics. Bubble diameter is 3 μm and drive field is 100-kHz triangular drive.

fact that the current margin of transfer-out operation can be improved up to 30 mA without bias margin loss by changing the pattern registration and shifting the conductor pattern a bit upward with reference to the permalloy pattern in Fig. 12a. This reduces the polarization effect at transfer-out gate half-disk pattern B. Generally speaking, it is effective to make permalloy patterns as large or as bulky as possible to reduce the polarizing effects of local fields, such as conductor current fields or bubble fields. This can reduce the effect of the local fields on the distribution of magnetization produced by the uniform rotating field. Figure 12c shows an improved version of the bi-directional transfer gate in which the transfer-in gate half disk pattern is enlarged, from 12 to 14 μm in width and from 9 to 11 μm in height by

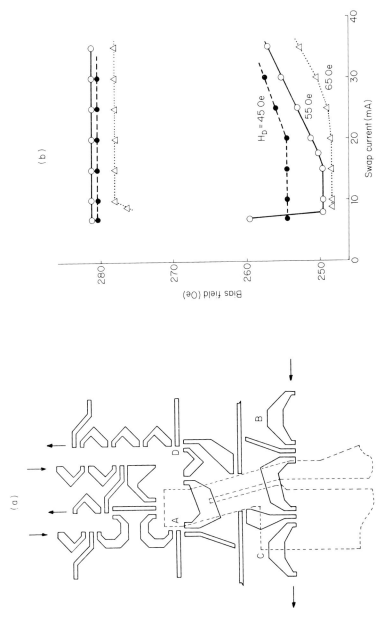

Fig. 13 (a) Swap-gate design with double period for an 8-μm period device. (b) Current margin curves at 200-kHz triangular drive fields of 45, 55, and 65 Oe.

removing the canted bar. The current margin of this design is improved sufficiently, as shown in Fig. 12d.

In true swap gates, it is more difficult to obtain a good current margin because a merge pattern (pattern E in Fig. 11c) must be located close to the transfer-in gate half disk pattern and bubbles to be transferred in are more easily stretched out to the merge pattern. As with the current margin of bi-directional transfer gates, low bias field margins are apt to be lost at high swap currents due to this stripe-out error, which is caused by the vertical field generated by the hairpin conductor (which reduces the bias field), as well as by the polarizing effect mentioned above. The true swap gate shown in Fig. 11c is an optimized version of the original design by Bullock *et al.* (1979b) and has a current margin from 15 to 35 mA, which is satisfactory. For 8-μm period devices, however, this problem is much more serious, and a design scaled down from Fig. 11c does not provide a satisfactory current margin. Furthermore, design optimization is very difficult because of spatial restrictions at the swap gate. An improved true swap gate is obtained by using the double period gate design concept to relax spatial restrictions. Figure 13a shows such a gate design for 8-μm period devices (Orihara *et al.*, 1979). The gate period is 32 μm and the major line pattern period is 16 μm, while the propagation period in the minor loops is held to 8 μm. Old bubble A and new bubble B are swapped precisely to positions C and D, respectively, after two cycles of field rotation. A satisfactory current margin is easily obtained with this design, as shown in Fig. 13b. Some other swap-gate designs for 8-μm period devices have also been proposed which employ double-period major line patterns (Bullock *et al.*, 1979a; and George *et al.*, 1981a).

3.4 Generator

For bubble generation, a nucleate generator (Nelson *et al.*, 1973) has been used instead of the disk generator which was used in the early development stages. Figure 14 shows the pickax nucleate generator design (Bonyhard and Smith, 1976), which is standard in 64-kbit to 256-kbit commercial devices with the gap-tolerant pattern design. When the top of the pickax pattern is magnetized, a rather high current (with polarity opposite to that of the replicator) is briefly applied through the hairpin conductor. The strong vertical field which reduces the bias field inside the loop can nucleate a bubble with the help of a magnetic pole at the top of the pickax pattern.

Figure 15 shows temperature dependences of the minimum and maximum generator currents for a pickax-type nucleate generator used in a 3-μm bubble, 16-μm period device (Matsuyama *et al.*, 1979). The minimum

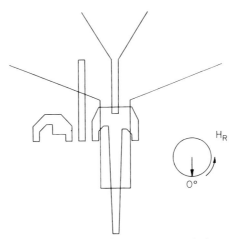

Fig. 14 Pickax nucleate generator design.

generator current is determined by the anisotropy field of the bubble material and by wafer processing (Guldi *et al.*, 1981; Johnson *et al.*, 1981), rather than by details of pattern design. Its linear temperature dependence, $-0.62\%/°C$, is regarded as reflecting the temperature dependence of the anisotropy field. On the other hand, the maximum current is far more dependent on temperature ($-1\%/°C$) and is determined by the pattern design and current pulse shape rather than by anisotropy. The generator

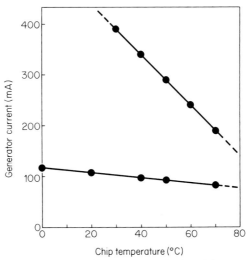

Fig. 15 Temperature dependence of current margin for pickax nucleate generator in the worst condition (3-μm bubbles, 100-kHz drive).

current is limited at the high end by the generation of extra bubbles in sparse data patterns at low bias fields, while it is limited at the low end by inability to nucleate bubbles. To make things worse, it is difficult to define the maximum current during the short measurement interval because the curve of error rate versus generator current is not sharp enough. In Fig. 15, data is plotted for an error rate of less than 10^{-8}.

As shown in Fig. 15, it is difficult to obtain a satisfactory current margin over a chip temperature range of 0–70°C or more. In practical cases, measures are adopted to guarantee stable operations. These include temperature tracking of the current (Naden et al., 1976) and use of auxiliary pulses in the same manner as replicator pulses (Matsuyama et al., 1979). Different generator designs have been proposed to solve this problem, such as combination of seed bubble nucleation and replication using a permalloy disk (Davies et al., 1980) or a small re-entrant loop with a pickax generator–replicator (Orihara et al., 1979). In these designs, bubble nucleation with large current is done only for generation of seed bubbles when the devices are initialized at a room temperature; data generation is performed by replication which does not have inherent temperature dependence and which requires less current than nucleation. Although the replicator-type generators provide better performances, users do not like the somewhat complicated initialization procedure.

When the bubble diameter is reduced, the generator current margin at high temperature is not improved (at least, judging from the poorer results obtained in a pickax generator of scaled design and using 1.5-μm diameter bubbles). Analysis of extra bubble generation through quasi-static (Matsuyama et al., 1979) and stroboscopic observation (Suzuki and Humphrey et al., 1980) indicates that when generator current is applied, bubbles are not only nucleated but stretched along the hairpin conductor, filling the inside of it. After the current terminates, the stripe shrinks back to the top of the pickax pattern, leaving extra bubbles around it which sometimes enter vacant positions according to data patterns written. Further, it is observed that the bottom of the nucleated stripe begins to go around the pickax head before the current terminates, and the hooked stripe seems to split more easily when it shrinks. According to these observations, it is concluded that the magnetic pole of the generator permalloy pattern must be localized while the current is being applied, and extra surrounding magnetic poles must be removed in order to improve the maximum generator current at high temperatures. Figure 16 shows an improved generator design for 8-μm devices which takes the above into consideration. In this design, called a hammer generator, a magnetic pole is formed only at the edge of the pattern during current application, and a stripe stretched along the conductor can shrink back to the edge in a stable manner. Figure 17 shows

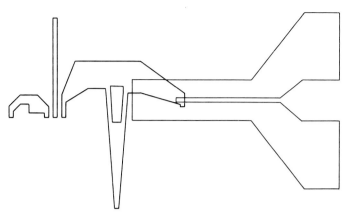

Fig. 16 Improved nucleate generator design: hammer generator.

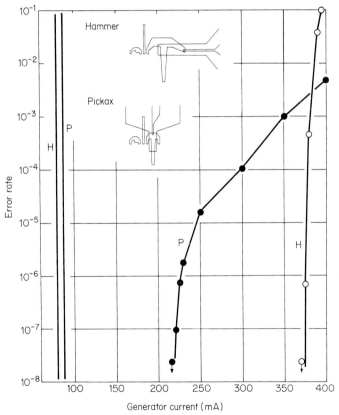

Fig. 17 Error rate curves of the minimum and maximum generator current in the worst condition: 70°C $(10\,000\,000)^n$ data pattern, and low bias field (100–kHz drive, 200-nsec pulse width, 1.9-μm bubbles).

Field Access Permalloy Devices 161

the error rate curves for minimum and maximum generator currents in the new generator compared with those for the pickax generator under the worst conditions: 70°C chip temperature, low bias field, sparse data pattern $(10\,000\,000)^n$ at 100 kHz with a 200-nsec generator pulse width. Error rate is defined as the number of errors divided by the number of read-out bits including zeros. As mentioned above, the error rate curve for the pickax generator is not sharp; in other words, its threshold is broad or vague. To determine the real threshold, generator must be tested at an error rate below 10^{-6}. On the other hand, the error rate curve with the improved design is very sharp, and the maximum current is improved remarkably. Thus, satisfactory generator current margin is expected over a wide temperature range without the temperature tracking.

3.5 Detector

Of the various detection methods proposed in the early stage of developments, so-called thick-film detectors (Strauss *et al.*, 1972) (which utilize the same permalloy film as propagation patterns) have been used in all commercial devices so far. Figure 18 shows typical examples of the design for a serpentine detector (Bobeck *et al.*, 1973) and a Nelson-type detector (Nelson, 1977). These chevron detectors are incorporated with chevron stretchers, and their resistances change because of a magneto-resistive effect caused by fields from stripe domains. Output signals are obtained as voltage changes across the detectors by applying dc current to them. Examples of waveform are shown in Fig. 19 for both of the detectors; voltage change due to the rotating field is cancelled by using a dummy detector.

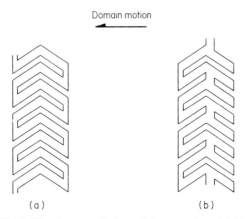

Fig. 18 Thick-film detector designs: (a) serpentine, (b) Nelson-type.

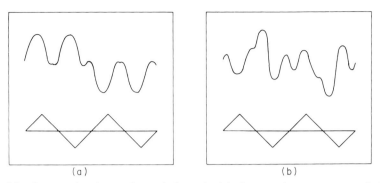

Fig. 19 Output signal waveforms balanced with dummy detector arranged side-by-side and drive field waveforms: (a) serpentine and (b) Nelson type.

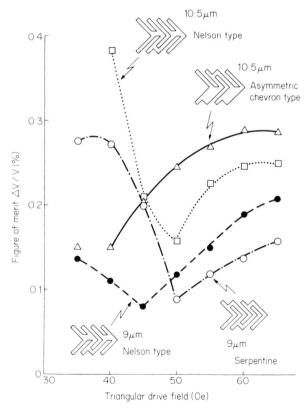

Fig. 20 Normalized output voltages versus drive field for various detector designs (1.5-μm bubbles, 100-kHz triangular drive).

The requirements for the chevron detectors are a good signal-to-noise ratio over the operating drive and bias field range, as well as sufficient stripe propagation. Magneto-resistive sensitivity of the detectors depends on permalloy film properties (magneto-resistive coefficient, coercive force, and magneto-strictive coefficient), as well as pattern shape and film thickness. The coercive force must be as low as possible because high coercivity increases the drive field required for propagation, not only at the detector itself but also in other parts of the device. A slightly Ni-rich film provides a high detector sensitivity with a low film coercive force, because its negative magnetostriction induces anisotropy longitudinal to the chevron bars, which acts to increase the sensitivity (Sakai et al., 1976). The Ni-rich content also yields a higher magneto-resistive coefficient. A problem encountered when the chevron detectors are scaled down for 8-μm period devices is occurrence of noise at higher drive fields, which tends to limit operational stability of the device at low drive fields. Although the cause of the noise is not fully understood, it is associated with domain switching in the detector chevron patterns and (unlike noise caused by inductive or capacitive coupling with drive coils) cannot be cancelled by balancing with a dummy detector. This switching noise occurs at the drive field where the magnetization mode of the chevrons changes and the ω–2ω transition of the output waveform (Bobeck et al., 1973; West et al., 1976) occurs; this drive field generally increases as the pattern width is reduced.

Figures 20 and 21 show output voltages and noise characteristics, respectively, of different chevron detector designs for detecting 1.5-μm bubbles. The composition of the permalloy film used here is 84% Ni, 16% Fe, the thickness is 3700 Å, and the coercive force is about 1.5 Oe. The output voltage and noise level are normalized by dc voltage across the detector. As shown in Fig. 21, the noise level is still rather high with a drive field of 50 Oe and does not disappear even up to 55 Oe for the serpentine detector which is half scaled from a detector for 3-μm bubbles, in which switching noise disappears when the field is below 35 Oe. On the other hand, the output voltage of the detector (shown in Fig. 20) is minimized when the drive field is reduced to about 5 Oe below the level at which switching noise disappears, then starts to increase with a different wave form (2ω to ω transition) when the drive field is reduced further. With these characteristics, this serpentine detector cannot be used at drive fields below 55 Oe.

Relationships between the pattern shape (including film thickness) and the field level at which switching noise disappears or the magneto-resistive sensitivity are very complicated and must be optimized through compromise with the stripe propagation performance by cut-and-try experiments. For example, a Nelson-type detector whose design is different from the serpentine detector only in the manner in which chevron bars are shorted exhibits

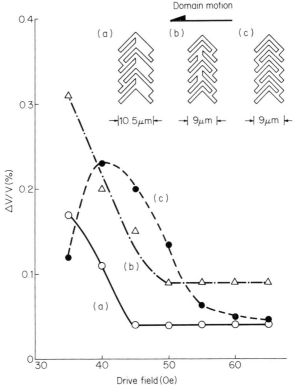

Fig. 21 Normalized noise levels versus drive field for various detector designs (1.5-μm bubbles, 100-kHz triangular drive).

better results, both for field level at which noise disappears and for sensitivity, while one with a column width increased to 10.5 μm provides higher detector sensitivity. Far better results are obtained for both characteristics by shorting half of the chevron bars in the 10.5-μm column period. The field level at which switching noise disappears seems to be reduced by reducing the shape anisotropy of the chevron bar, although this generally results in a poorer stripe propagation margin with low bias fields. The modified Nelson detector shown in Fig. 20 exhibits good propagation performance at low drive fields, as does the similar shape stretcher shown in Fig. 4. However, it requires the addition of some standard chevron patterns to the ends of the column to prevent stripe domains from stretching out to detector leads. The modified Nelson type detector provides an output voltage of about 8 mV and a signal-to-noise ratio of about 10 with 500 chevrons (resistance: 1.2 kΩ), 3 mA bias current, and a 45-Oe–65-Oe drive field at the midpoint of the bias field range.

4 OPTIMIZATION OF PROPAGATION

This section discusses optimization of various parameters to obtain satisfactory propagation characteristics in minor loops, such as bubble diameter, pattern design, pattern period, spacing, and so forth. The discussion concerns optimization for an 8-μm period device.

4.1 Bubble diameter and pattern design

The optimum nominal bubble diameter d for a fixed propagation period p was given as $d = \frac{3}{16}p$ for the T-bar propagation pattern by Bobeck and Della Torre (1975). This rule has also been applied with 16-μm period gap-tolerant propagation patterns, and a 3-μm nominal bubble diameter is used in commercial devices. For 8-μm period devices, however, 2-μm bubbles have been used widely in development, instead of the 1.5-μm bubbles predicted by the rule.

For minor-loop propagation patterns, asymmetric chevron patterns have been widely employed in 8-μm period devices as well as in 16-μm period devices. In most cases, however, the patterns are slightly modified from the original design shown in Fig. 1 because the acute angles at the edges make it difficult to produce with the usual photo-mask generation systems, which build patterns out of small rectangles.

When evaluating minor loop propagation, it is important to test chips under conditions simulating the worst condition actually encountered in memory devices, such as various data patterns, start–stop operations, application of a dc in-plane field for holding bubbles during start–stop operations, etc. Among the data patterns, the worst case must be chosen to sufficiently include bubble–bubble interaction effects. Figure 22 shows a comparison of minor-loop propagation margins for different pattern designs and bubble diameters (Yanase et al., 1980). The designs of the asymmetric chevron and half disk patterns used are shown in Fig. 23. The pattern periods and the track periods are both 8 μm, while the pattern gaps are 1 μm. The composition of the LPE (liquid-phase epitaxy) films is $(YSmLuCa)_3 (GeFe)_5 O_{12}$, and the film parameters are 1.6 μm and 1.9 μm in stripe width, 1.7 μm and 1.9 μm in thickness, and 510 G and 430 G in $4\pi M_s$, respectively. The permalloy pattern thickness is 3700 Å and the garnet-to-permalloy spacing is 4800 Å. In Fig. 22, the bias margins of straight propagation in the minor loops are measured at 100 kHz with start–stop operations for the worst-case bubble data pattern. A dc in-plane field of 6 Oe is applied at right angles to the direction of propagation. This holding field shifts the rotating drive field locus in one direction and diminishes the drive field at the pattern

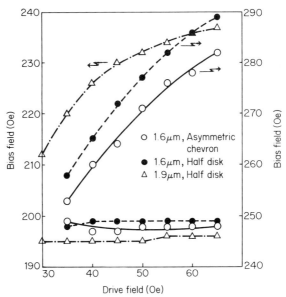

Fig. 22 Comparison of propagation margins for asymmetric chevron and half-disk tracks and for 1.6-μm and 1.9-μm diameter bubbles (100-kHz triangular drive).

gaps in either the going or returning tracks where the bias margins are determined. The bias margins in the figure are those of the disadvantaged tracks. For 1.6 μm bubbles, the half-disk pattern tracks show better bias margins at high bias fields than the asymmetric chevron pattern tracks. This difference is not observed for propagation of isolated bubbles without the holding field. Also judging from the similar results obtained for 1.9-μm bubbles in 8-μm period tracks and for 3-μm bubbles in 16-μm period tracks, propagation for the chevron track seems, in general, to be less stable than for the half-disk track at low drive fields with full bubble loading and application of a dc in-plane field. The superiority of the half-disk pattern is attributed mainly to increased permalloy volume, which could enhance the bubble drive force at the pattern gaps. The result, however, does not mean that the pattern designs called asymmetric chevron are always inferior to half-disk patterns, since the asymmetric chevron used here is slightly different from the original design mentioned above.

When the pattern period and gap are fixed, small bubble films would give lower high-bias thresholds because the bubbles at high bias fields would be too small to propagate across the pattern gaps; large bubble films would give higher low-bias thresholds because the bubbles at low-bias fields would be so large that they would easily stripe-out to nearby attractive magnetic poles. As shown in Fig. 22, the 1.9-μm bubble film provides better bias

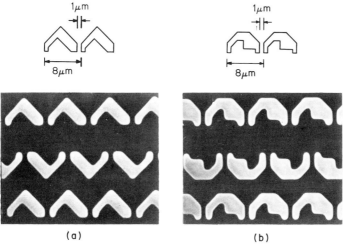

Fig. 23 Designs of 8-μm period: (a) asymmetric chevron and (b) half-disk patterns.

margins at low drive fields in the half-disk propagation track than the 1.6-μm bubble film, and without deterioration at the low-bias threshold. This is attributed to differences in saturation magnetization, as well as in bubble diameter. Bubble–bubble interaction affects the high-bias thresholds, but significant differences are not observed between the two films; the difference between the threshold with isolated bubbles and that with fully loaded bubbles averages a few oersteds for both films. Although the possibility of further margin improvements with still larger bubble films might be foreseen from these results, the 1.9-μm nominal diameter seems optimum considering the satisfactory results shown in Fig. 22 and propagation margins at turns or functions where bubbles might strip-out more easily because of complicated pattern layouts.

4.2 Pattern period, gap, and spacing

Bubble–bubble interaction and bubble drive force from the pattern might be changed by varying the propagation period while keeping the pattern gap constant for a fixed bubble diameter. Figure 24 shows the period dependence of the straight propagation margin. The nominal bubble diameter is 1.9 μm, the propagation track consists of a half-disk pattern with a 1-μm gap and 7-μm track period, and other conditions are the same as in the case of Fig. 22. As the period is reduced from 9 to 6 μm, the minimum drive field increases significantly, and below 7 μm it is difficult to obtain sufficient bias margin with a drive field of 45 Oe.

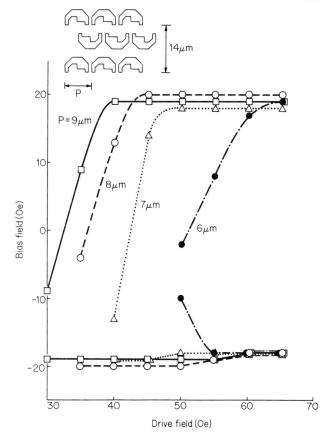

Fig. 24 Period dependence of straight propagation margin curve in half-disk tracks with 1-μm gap and 7-μm track period for 1.9-μm bubbles.

On the other hand, when the pattern gap is increased from 1 μm with the pattern period kept constant for a fixed bubble diameter, deterioration of the propagation margins is easily predicted because of weaker drive forces at the pattern gaps. Figure 25 shows pattern gap dependence of the 8-μm period half-disk propagation margin for 1.9-μm diameter bubbles. Conditions for processing and measurement are the same as in Fig. 24, but the bubble pattern is an isolated one. The bias margin deteriorates as predicted when the pattern gap is increased to 2 μm. It is interesting, however, that the minimum drive field required does not vary, which differs from the period dependence shown in Fig. 24.

In Fig. 24, an increasingly higher minimum drive field is required as the period is reduced. This is mainly because of the stronger bubble–bubble

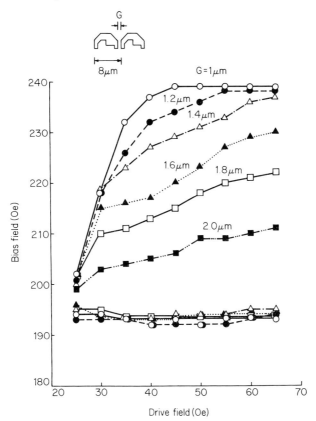

Fig. 25 Gap dependence of 8-μm period half-disk propagation margin curve for 1.9-μm bubbles.

interaction, rather than deterioration of the bubble drive force from the reduced-size propagation pattern. This is deduced by comparing the data in Figs. 24 and 25. First, comparison of the margin curve for fully loaded bubble propagation in the 8-μm period track shown in Fig. 24 and that for isolated bubble propagation in the 1-μm gap track shown in Fig. 25 shows that the bubble–bubble interaction increases the minimum drive field by about 5 Oe and does not affect the bias margins at high drive fields. Second, the reduction in half-disk pattern size cannot explain the considerable change in minimum drive fields shown in Fig. 24, considering that there is no difference in the minimum drive fields shown in Fig. 25. The half-disk size of the 7-μm period track in Fig. 24 is the same as that of the 2-μm gap track in Fig. 25. On the other hand, the linear dependence of bubble diameter on the bias field can explain the gap dependence shown in Fig. 25. The minimum

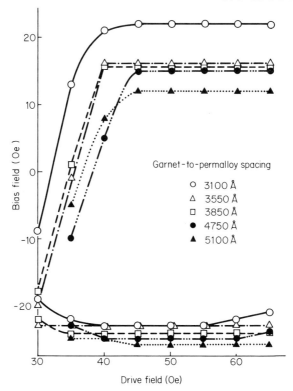

Fig. 26 Spacing dependence of 8-μm period half-disk propagation margin curve for 1.9-μm bubbles.

diameter d of a bubble which can propagate across the gap is determined by the gap size g; roughly speaking, $d = g + w$ (w is the bar width of the pattern edge) (Almasi and Lin, 1976). As the bubble diameter under the pattern edge is linearly dependent on the bias field, the high-bias threshold (the bias field which provides the minimum diameter) varies approximately linearly with the gap size as shown in the figure. Further, the minimum drive field does not change as long as the bubble diameter at low bias end is larger than the minimum bubble diameter determined by the gap size.

Spacing between the bubble film and the permalloy pattern is another parameter to be optimized for propagation. A small spacing is expected to result in strong coupling between propagation patterns and bubbles and to reduce the minimum drive field. Spacing which is too small, however, is predicted to strengthen the self-stabilizing effect at the same time and to increase the minimum drive field (George et al., 1974). The minimum spacing is also limited by the restrictions of wafer processing; smaller spacing might increase the stress effect of overlay patterns or make it difficult to obtain smooth step coverage at conductor pattern.

Figure 26 shows spacing dependence of the straight propagation margin in 8-μm period half-disk tracks with 1-μm gaps and 1.9-μm bubbles. The conditions of measurement are the same as in Fig. 22. The spacing is varied from 3100 to 5100 Å by changing the thickness of the planarizing resin layer between the conductor and permalloy layers (explained later). Within the range covered by the figure, the smaller the spacing, the better are the minimum drive field and the bias margins at high drive fields. The smaller spacings improve the efficiency of bubble driving and increase the collapse field of the bubbles under the permalloy patterns. At high drive fields, high-bias thresholds are generally limited by start–stop operations; the thresholds are regarded as representing the collapse field of bubbles trapped under the permalloy patterns. Without the restrictions of the wafer processing, even smaller spacings seem to act to improve propagation margins.

The propagation margin is not strongly dependent on the thickness of the permalloy pattern, provided that the thickness is great enough to prevent magnetic saturation in the required drive field range. Although the propagation margin tends to improve as the pattern thickness is increased, the maximum thickness is limited by deterioration in detector performance.

4.3 Effect of crystallographic orientation

The effect of crystallographic orientation on propagation margin is well known in ion-implanted devices. This is also observed in permalloy devices (Yanase et al., 1980). Figure 27 shows the propagation margins of half-disk tracks without dc inplane field for 1.9-μm bubbles. Other conditions are the same as in Fig. 22. As shown in Fig. 27a, the propagation margin of track A, which is aligned along crystallographic orientation ($\bar{1}\bar{1}2$), is superior to that of track B, which is aligned along ($11\bar{2}$). This alignment is defined as 0° orientation, where ($\bar{1}\bar{1}2$) is oriented to the 0° direction of the rotating drive field. On the other hand, in 180° orientation, where ($\bar{1}\bar{1}2$) is oriented to the 180° direction of the drive field, the propagation margin of track B is superior to that of track A, as shown in Fig. 27b.

When the dc in-plane field is applied in the 90° direction of the drive field, the bias margin difference shown in Fig. 27a is enlarged as shown in Fig. 28a; upper thresholds of bias margins for track A are improved, while those for track B deteriorate. The same behavior with respect to the dc in-plane field is observed for the 180° orientation as shown in Figs. 27b and 28b. In this case, the bias margin difference is reduced. Therefore, the behavior of the threshold on the dc in-plane field does not depend on crystallographic orientation. As

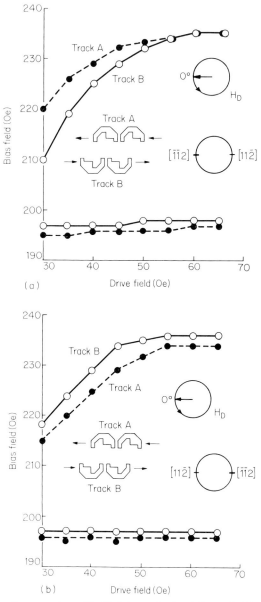

Fig. 27 Propagation margins of half-disk tracks for 1.9-μm bubbles without dc in-plane field in (a) 0° orientation and (b) 180° orientation.

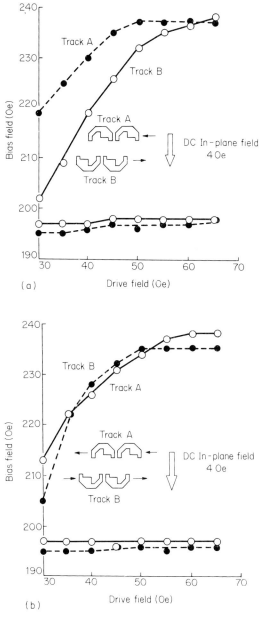

Fig. 28 Propagation margins of half-disk tracks for 1.9-μm with dc in-plane field of 4 Oe in (a) 0° orientation and (b) 180° orientation.

shown in Fig. 28b, compensation for the bias margin difference can be obtained by orienting crystallographic orientation ($\bar{1}\bar{1}2$) to the 180° direction of the rotating drive field (the propagation direction for track B), and applying the dc in-plane field along the 90° direction of the drive field. Similarly, in 0° orientation, compensation for the bias margin difference can be obtained by applying the dc in-plane field along the 270° direction of the drive field.

The results shown in Fig. 27—that is, the difference in propagation margins between tracks A and B—are reversed by changing the crystallographic orientation by 180°; this shows that the difference is caused by in-plane anisotropy of the garnet film. This appears to be the same as in the ion-implanted devices, where anisotropic propagation margins are caused by threefold symmetric charged walls (Lin *et al.*, 1978). In this case, however, the difference in bias margins for propagation direction is supposed to be related to the threefold symmetric response of closure domains associated with bubbles to in-plane fields (Uchishiba *et al.*, 1977). As shown in Fig. 28, the upper thresholds of the bias margins for track A are improved by the application of the dc in-plane field, while those for track B deteriorate. The in-plane field directly increases the drive field at the pattern gaps in track A (where the margin thresholds are determined), while it reduces the drive field in track B. Thus, the upper thresholds of the bias margins for track A are improved while those for track B are degraded. The direction of the in-plane field can be chosen without regard for the garnet film orientation. The margin threshold difference caused by in-plane anisotropy of the film can be cancelled by the difference produced by the dc in-plane field, as shown in Fig. 28b.

As the propagation margin difference between directions is related to the threefold symmetric in-plane anisotropy field of the garnet film, it is expected to be dependent on ion-implantation conditions for hard bubble suppression and strength of the vertical anisotropy field of the film. In fact, the difference is observed to decline as the ratio of the in-plane anisotropy field to the vertical anisotropy field decreases.

4.4 Conductor crossing

Propagation margins deteriorate considerably at conductor crossings, especially in high-density devices. The minimum drive field for propagation at gate patterns or major loops with conductor crossings is 5–10 Oe higher than without conductor crossings, because of bubble trapping at the conductor edges. This bubble trapping is attributed to several different causes. The first is metallization stress of the overlay patterns (Dishman *et al.*, 1974). Local stress fields from the metal edges perturb the energy of domain walls through magneto-striction, which induces either attractive or repulsive potentials for bubbles at the conductor edges. The second is discontinuity of pattern magnetization at sharp conductor edges due to poor step coverage. The drive field induces spurious magnetic poles at the conductor edges, which obstruct bubble propagation. The third is an effect of the bias field. Even with smooth step coverage, static bias field distortion can result because of the component of the field along the step (Roman *et al.*, 1979). Finally, difference in spacing across the conductor step might affect the propagation margins.

These causes have a greater adverse effect on propagation margins as the propagation period and the bubble diameter are reduced. The conductor thickness cannot be reduced because of the need to prevent electromigration, as well as to avoid increasing gate resistance, so the aspect ratio of the conductor cross-section will increase. Further, the bias field and drive field increase as the bubbles are made smaller.

To solve the step coverage problem, various kinds of planarizing processes have been proposed. These are represented by two processes: the planar process and the semi-planar process. In the planar process, the surface of the wafer is made completely flat after the conductor patterns are delineated, by embedding them with SiO_2 (Reekstin and Kowalchuk, 1973) or SiO (Orihara *et al.*, 1978) using lift-off techniques. Permalloy patterns deposited on this flat surface should not have magnetic discontinuities or vertical components. In actual cases, however, it is very difficult to obtain perfect flatness because of limitations in process technology, and small (but sharp) steps or grooves are left along the conductor edges; these give rise to the magnetic discontinuities. Further, in this process the spacing between the garnet film and the permalloy patterns cannot be reduced to less than the conductor thickness. This restriction limits improvement of the minor-loop propagation margin by minimizing spacing. On the other hand, with the semi-planar process, the slope and height of the conductor steps are reduced by spin-coating with some kind of resin (Sugita *et al.*, 1979; Majima *et al.*, 1981; George *et al.*, 1981a). In this case, the conductor steps are smoothly covered and the permalloy patterns are easily deposited without magnetic discontinuities, although the vertical component at the slope might cause the bias field effect. In contrast with the planar process, spacing can be reduced to some extent regardless of the conductor thickness to improve the minor loop propagation margin.

Figure 29 shows layer structures and scanning electron microscope SEM photographs of 8-μm period devices fabricated by the SiO lift-off planar process and the resin semi-planar process (Majima *et al.*, 1981). The thickness of the conductor patterns is 4200 Å in both cases. With the SiO lift-off

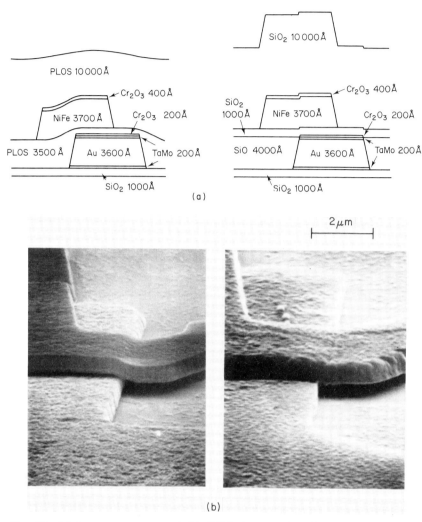

Fig. 29 (a) Layer structures and (b) SEM photographs of 8-μm period devices fabricated with the SiO lift-off planar process (right) and the resin semi-planar process (left).

process, the steps at the edges of the conductors are planarized with a 4000-Å SiO layer. A small 200-Å step is used for pattern registration in this case, but this step height should be taken into consideration as a planarizing variation in processing when it is not required for registration. In this planar process, spacing between the garnet film and the permalloy patterns is 6000 Å at the conductors; in the resin process, the thickness of the resin is 3500 Å and the

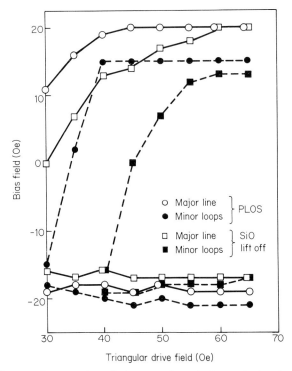

Fig. 30 Propagation margins of planar and semi-planar devices in 14-μm period major-line tracks with conductor crossings and 8-μm period minor-loop tracks for 1.9-μm bubbles.

spacings are 4500 Å in the area without conductors and 6950 Å at the conductors.

Propagation margins of both devices were measured in 14-μm write major lines with swap gates shown in Fig. 13, as well as in the 8-μm period half-disk minor-loop tracks. Figure 30 shows the results for both the planar and semi-planar process. The resin device even exhibits a slightly better margin than the SiO lift-off devices for major-line propagation. This means that the sharp 200-Å steps at the conductor edges can affect the propagation margin while the smooth 2450-Å steps of the resin process do not affect it at all. Available data indicates that the magnetic discontinuity effect is more important than the bias-field effect. However, both margins are acceptable since they are much better than the minor loop propagation margins. As for minor-loop propagation, the resin device has considerably better bias margins at low drive fields than does the SiO lift-off device. This difference is attributed entirely to spacing. In the minor-loop area, the spacing of the resin device is 1500 Å less than that of the SiO lift-off device. A 6000-Å spacing seems to be too great for the 8-μm period pattern to attain an acceptable propagation margin. On the other hand, the propagation margins of the major line are sufficient in both devices, although spacing at the conductor is greater than 6000 Å. This shows that the propagation margin of the 14-μm period permalloy pattern in the major line is not so critically dependent on spacing as the 8-μm period propagation margin in the minor loops.

The resin semi-planar process, together with the double-period or enlarged propagation pattern design, can solve most conductor crossing problems and is to be widely employed in devices with periods of 8 μm or less. For the 8-μm, 1-Mbit device, another planar process (called the top-down process) is proposed (Bullock *et al.*, 1979a). As the conductor is left under the permalloy patterns after processing, the magnetic discontinuities of the lift-off planar processes are not encountered. However, it is expected that the large spacing of this device will limit its applications.

5 LOOP ORGANIZATION

The arrangement of propagation tracks together with the various functions is an important factor in determining the performance of memory devices (access time, data rate, data integrity, ease of read–write control, and so forth). The simplest arrangement is the single-loop organization consisting of a long re-entrant propagation track, a generator, a replicator, and a detector. As this arrangement is relatively simple to design and easy to control compared with the multi-loop organizations explained below, 64-kbit level devices with this organization have been used for practical applications (Bobeck and Danylchuk, 1977; Yanase *et al.*, 1978). However, access time is very long; for example, an average of 0.64 sec is required for a 64-kbit capacity with 100-kHz rotating field cycle. This time is proportional to the capacity, and even a single defect makes the whole device useless. This means that the single-loop organization is not suitable for large-capacity devices. On the other hand, multi-loop organizations can solve these problems of the single-loop organization and are suitable for large-capacity devices. One early proposal is a major–minor-loop organization (Bonyhard *et al.*, 1970) which consists of many small propagation loops (minor loops), an access loop with a generator and a detector (a major loop), and transfer gates to connect them. A device with some defective minor loops can be used by replacing them with redundant minor loops by means of control electronics. The average access time is dramatically reduced (to the order of milliseconds) because of the short length of the minor loops. Although 64-kbit to 256-kbit devices with this loop organization have been developed for

Field Access Permalloy Devices 179

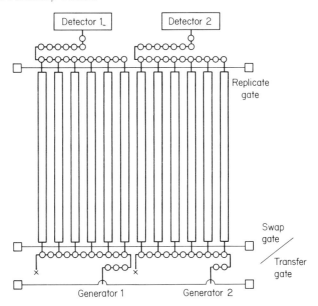

Fig. 31 Even–odd block-replicate chip design.

practical uses (Hiroshima *et al.*, 1976; Naden *et al.*, 1976; Orihara *et al.*, 1978), the control required to rewrite bubbles in the minor loops which have been transferred out to the major loop for read operations and to guarantee data integrity against power failures is bothersome.

A block-replicate organization is widely employed for commercial 256-kbit devices (Ypma *et al.*, 1976; Bonyhard and Smith, 1976). Figure 31 shows a schematic example. Minor loops are connected at the top to read major lines through replicate gates, and at the bottom to write major lines through transfer gates or swap gates. During read operations, the bubbles making up a page stored in the minor loops are replicated simultaneously at the top of each minor loop; as the bubbles in the read line need not be written back into the minor loops, they are annihilated or disposed of outside a guard rail. In Fig. 31, the minor loops are divided into two blocks, even and odd. In these, the number of propagation steps from the first loop to the detector differs by one step. As the minor-loop period is twice the major-line period, replicated bubbles occupy every other bit position in the major lines and are detected every other cycle. Due to the difference in steps, the bubbles from each block can make up outputs in every cycle at sense circuit output. On the other hand, in write operations the same bubble data is generated in each of the write lines; the even bits of this data are transferred into the even block minor loops (and vice versa) due to the one step difference from the generator to the last minor loop between the blocks. The odd bits in the even write line (and vice versa) are disposed of outside the guard rail. When the transfer gates are used for write gates, the page in the minor loops to be written should be erased beforehand by transferring it out to the read lines via the replicate gates. In the case of the swap gates, an erase operation is not required and new data and old data are swapped at one gate operation. In one version of the block-replicate organization, the even and odd read lines are merged and one detector reads bubbles during every cycle (Bullock et al., 1979b), although at the expense of some degradation in signal-to-noise ratio.

Another point to be explained concerning loop organization is on-chip measures for housekeeping data: i.e., page address marks and bad-loop maps. In one case, one of the minor loops is allotted to a marker loop and a mark (bubble) is stored in it for the first page. The bad-loop map is stored in a PROM (Programmable Read Only Memory) outside the bubble device. In another case, the first page mark is stored in a minor loop and the bad-loop map is stored in the first page. In the bad-loop map, the existence of a bubble indicates a good loop and a vacancy indicates a bad loop (Kita et al., 1980). In a recent technique, a special loop comes to be used for storing the first page mark and the bad-loop map (Bullock et al., 1979b). This loop is called a redundancy or boot loop; it possesses its own gate leads and is controlled separately from the storage loops. This organization is easy to handle, and the possibility of destroying housekeeping data through mishandling before system operation is expected to be very low compared with other schemes.

The first consideration in designing loop organizations for high-density devices with propagation periods of $8\,\mu m$ or less is that of relaxing the spatial restrictions at the gates to reduce the drive field, without degrading device performances or increasing chip size. Considerations of gate design indicate that the bias margins with low drive fields can be improved remarkably and the drive field can be reduced to the level of the drive field for 16-μm period devices if the gate period is increased to twice that of the

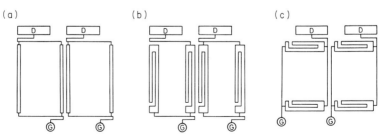

Fig. 32 Loop organization for relaxing spatial restrictions at the gates.

conventional design shown in Fig. 31. This is easily achieved by connecting two neighbouring loops and folding the loops as shown in Fig. 32b. However, the minor-loop length is doubled and access time deteriorates if the chip is assumed to be approximately square, which is preferable from the point of view of chip fabrication and drive coil design. Further, the data rate on the major line is reduced to one-fourth that of the rotating field cycle. These problems of the double-period gate organization can be solved by changing the minor-loop layout and employing double-period major lines as shown in Fig. 32c (Bonyhard, 1979; Orihara et al., 1979). This loop organization is the full equivalent of the conventional even–odd block-replicate organization shown in Fig. 31 in terms of chip size, read–write data rate, access time, and page size (see Table 1). The gate designs described in Section 3 can be

Table 1 Comparison of loop organizations in Fig. 32.

Characteristic[a]	Loop organization		
	(a)	(b)	(c)
Number of minor loops	$\sqrt{N}/2$	$\sqrt{N}/4$	$\sqrt{N}/2$
Minor loop length	$2\sqrt{N}$	$4\sqrt{N}$	$2\sqrt{N}$
Average access time	$1.25\sqrt{N}/f$	$2.25\sqrt{N}/f$	$1.25\sqrt{N}/f$
Data rate	f	$f/2$	f

[a] N is chip capacity, f is drive frequency, and the average access time is the access time for the middle bit of an average page without the latency time from the first loop to the detector.

incorporated into this loop organization, making it possible to reduce the drive field and relax the requirements for pattern registration, conductor reliability, and so forth. A similar loop organization with folded minor loops is employed in the enhanced density design concept (Dimyan and Hubbell, 1979). In this design concept, however, function designs are standard 16-μm period ones for 3-μm bubble devices, and the minor-loop cell size is reduced to $10.5 \times 10.5\,\mu$m. Other loop organizations proposed for 8-μm, 1-Mbit devices utilize only the double-period major lines, improving the bias margins of swap gates and the data rate (Bullock et al., 1979a; George et al., a).

As the bubble memory is essentially a kind of shift register, access time deteriorates with increased chip capacity. The access times for the chip with the loop organization shown in Fig. 31 are as follows. Access time for the first bit of an average page is

$$T_{AF} = \frac{L}{2} \times \frac{1}{f} + \alpha = \frac{\sqrt{N}}{f} + \alpha.$$

Table 2 Average access time versus chip capacity in the even–odd block organization.

Characteristic[a]	Chip capacity N (bits)			
	64 k	256 k	1 M	4 M
Number of minor loops n	128	256	512	1024
Minor loop length L (bits)	512	1024	2048	4096
Average access time for the first bit, T_{AF} (msec)	3.6	6.1	11.2	21.5
Average access time for the middle bit, T_{AM} (msec)	4.2	7.4	13.8	26.6

[a] Drive frequency is 100 kHz and the latency time (α) assumed is 1 msec.

Access time for the middle bit of an average page is

$$T_{AM} = \left(\frac{L}{2} + \frac{n}{2}\right) \times \frac{1}{f} + \alpha = \frac{5\sqrt{N}}{4f} + \alpha.$$

In these equations, N is chip capacity, L is minor loop length, n is number of minor loops, f is drive frequency, and α is a latency time for the propagation from the first loop to the detector. To make the chip approximately square, $n = L/4$ and $L = 2\sqrt{N}$ are assumed. In Table 2, T_{AF} and T_{AM} are shown for chip capacities of 64 kbit to 4 Mbit with a 100-kHz drive frequency. To be more advantageous than magnetic disk memories, the access time of bubble memories must be held to less than about 10 msec; access times of more than 20 msec for a 4-Mbit device do not seem acceptable.

A direct approach to improving the access time, as well as the data rate, is to increase the drive frequency. Although the bubble memory chip itself could be operated rather easily at frequencies up to 400 or 500 kHz, problems such as high coil drive voltages or large power dissipation in the devices would limit the practical drive frequencies to less than 200 kHz. Here, an example of a loop organization with short access time is discussed, with 4-Mbit devices in mind. The loop layout shown in Fig. 32c makes it possible to divide a chip into more blocks with shorter minor loops, as shown in Fig. 33 (Orihara *et al.*, 1979; George *et al.*, 1981b). If m is the number of even–odd block pairs which make up the chip, the access times are as follows:

$$T_{AF} = \frac{\sqrt{N}}{mf} + \alpha,$$

$$T_{AM} = \left(\frac{1}{m} + \frac{1}{4}\right)\frac{\sqrt{N}}{f} + \alpha.$$

Field Access Permalloy Devices

Table 3 Access time and data rate for the multi-block organization.

Characteristic[a]	Chip capacity N (bits)					
	1 M			4 M		
	$m=1$[b]	$m=2$	$m=4$	$m=1$	$m=2$	$m=4$	$m=4$
Number of minor loops/m	512	512	512	1024	1024	1024	512
Minor loop length (bits)	2048	1024	512	4096	2048	1024	2048
Average access time for the first bit, T_{AF} (msec)	11.2	6.1	3.6	21.5	11.2	6.1	11.2
Average access time for the middle bit, T_{AM} (msec)	13.8	8.7	6.1	26.6	16.4	11.2	13.8
Data rate (kbit/sec)	100	200	400	1000	200	400	400

[a] Drive frequency is 100 kHz, the latency time (α) assumed is 1 msec, and each even–odd block pair is driven in parallel.
[b] Where m is number of block pairs.

Table 3 shows the access times versus m for chip capacities of 1-Mbit and 4-Mbit with a 100-kHz drive frequency. Significant improvement is provided by the multi-block organization, and access times of the 4-Mbit device can be kept within the acceptable range. Table 3 also lists data rates for access of block in parallel. The data rate is more important than the access time for large-capacity file memories, where the essential factor is how much data can be transferred to a central processing unit (CPU) in a short time. The data rate of a device can also be improved, as shown in Table 3, by the multi-block organization. The drawbacks would be increases in the number of function drive and sense circuits, as well as the number of chip leads. The

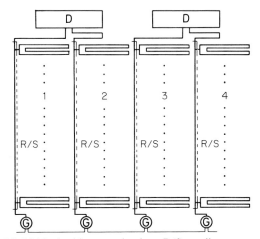

Fig. 33 Multi-block chip organization. R/S, replicate–swap gate.

development of integrated circuit (IC) support electronics, however, could solve these drawbacks, and the multi-block organization seems likely to become a standard loop organization for bubble memory devices with capacities of 4-Mbit or more (George et al., 1981b). Other approaches for improving the access time have been proposed; these include clockwise and counter-clockwise data shift in the minor loops (Yoshimi et al., 1978) and small cache loops connected to the main loops with swap gates (Kohara et al., 1981).

6 DESIGN AND CHARACTERISTICS OF A 1-Mbit CHIP

Based on the design considerations described in the previous sections, a full-scale 1-Mbit chip has been designed, fabricated, and characterized (Yanase et al., 1981; Majima et al., 1981). Outlines of its design and some of its characteristics are described below.

6.1 1-Mbit chip design

An even–odd organization for a 1-Mbit chip is shown in Fig. 34, where the loop organization is optimized from the standpoints of chip performance, pattern design, wafer processing, and packaging. Each block has 299 minor loops and a boot loop, both of which are 2053 bits in length. These are connected to a write major line with swap gates on the left side and to a read major line with replicate gates on the right side. Data is written by a generator at the bottom and read out by a stretcher–detector at the top. The minor-loop propagation tracks are of the 8-μm period half-disk type, and the track period is 7 μm. They are folded, and they allow the gate period to be doubled to 28 μm. The propagation period in the write and read major lines is 14 μm, and bubbles are read out from each block in every other cycle. The data transfer rate of the chip is 100 kbit/sec at a coil drive frequency of 100 kHz, while the access time is 11.2 msec for read out of the first bit of an average page. The boot loop has separate gate leads, although the gate designs are the same as those of the minor loops. Figure 35 shows a photograph of the 1-Mbit chip; the size is 9.1 × 9.9 mm. The chip is fabricated on a 1.9-μm bubble garnet film by the resin planar process. The composition of the film is $(YSmLuCa)_3(GeFe)_5O_{12}$, and typical parameters are 452 G in $4\pi Ms$, 1·85 μm in thickness, and 1·9 μm in stripe width. The chip fabrication process is as follows. First a 1000-Å SiO film is sputtered onto the substrate; then a 3600-Å Au film sandwiched between 200-Å TaMo alloy adhesion layers (Majima et al., 1981) is formed by evaporation in the same vacuum; next, conductor patterns are delineated by ion milling. Afterwards, a 3500-Å

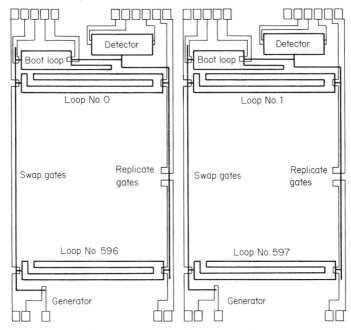

Fig. 34 1-Mbit chip organization.

ladder-type organosiloxane resin PLOS (Poly Ladder Organo Siloxane) film (Takeda et al., 1980) is spin-coated and cured, then a 3700-Å permalloy (84% Ni) film is deposited by evaporation, and propagation patterns are delineated by ion milling. Finally, a 1-μm PLOS passivation is spin-coated and cured. The Au conductor can be replaced by an Al-Cu conductor since current density is as low as that in the 16-μm devices (where their reliability has already been confirmed) because of the wide conductor patterns of the double-period gates. A schematic diagram is shown in Fig. 29a.

The conductor and permalloy layers are patterned using a 10:1 projection photolithographic system. Figure 36 shows SEM photographs of the swap gates, replicate gates, generator, and stretcher–detector. The swap-gate design is a half-disk version of the design of the true swap gate shown in Fig. 13. The replicate-gate design is one with the pickax pattern shown in Fig. 9, and is 13.6 μm in pickax width. The generator design is the nucleate generator shown in Fig. 16. The pattern width of the hairpin conductors is 4 μm, and the minimum feature size in the permalloy patterns is 1 μm in all cases, including the minor loops. The stretcher–detector consists of the 13.6-μm period Mt. Fuji stretcher shown in Fig. 6 and the 10.4-μm period asymmetric chevron detectors shown in Fig. 21. The number of chevrons in the stretcher

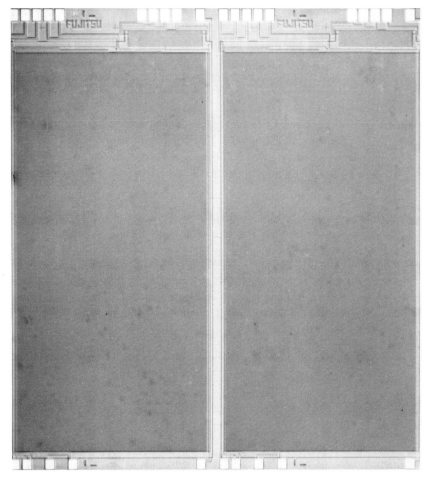

Fig. 35 Photograph of the 1-Mbit chip.

and detector is 500, and the number of stretcher columns is 18. The two detector columns are arranged side-by-side for differential connection to a sense circuit to minimizing noise in the detection of bubbles in every other cycle. Each active area is surrounded by a guard rail which expels bubbles to the outside, as shown in Fig. 35.

6.2 1-Mbit chip characteristics

The fabricated chip should be characterized for various parameters, such as bias field, drive field, function parameters, temperature, bubble loading,

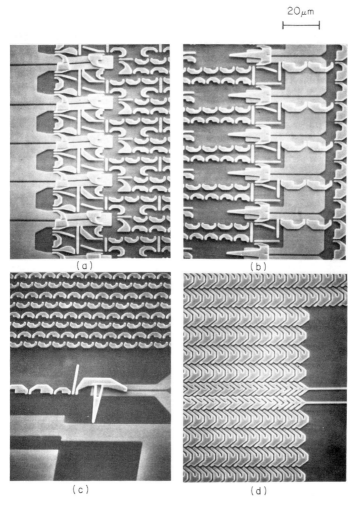

Fig. 36 SEM photographs of the 1-Mbit chip showing function designs: (a) swap gates, (b) replicate gates, (c) generator, and (d) stretcher–detector.

long-term operation, and so forth. Figures 37 and 38 show overall bias margins against the most critical function parameters; these are the swap current and the replicate pulse phase, respectively, and are shown at chip temperatures of 0, 33, and 90°C. The drive field is produced by a 100-kHz triangular wave and has a peak value of 55 Oe, and 32 pages are written and read out in the worst-case data pattern. As the replication operation is also dependent on the pulse amplitude, though to a lesser extent, the data shown in Fig. 38 were obtained by varying the amplitudes ±15% from the

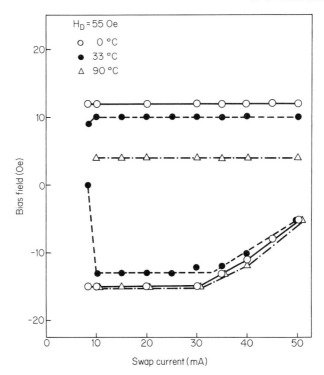

Fig. 37 Overall bias margin versus swap current at three chip temperatures (55 Oe peak field of 100-kHz triangular drive).

nominal values as shown in the inset. The high- and low-bias thresholds in the flat ranges are determined not by the gate operations but by minor loop propagation (especially at turns), except at 90°C where the output signal falls below 3.5 mV and limits the high-bias thresholds. The data in the figures shows a swap current margin of 20 mA ($\pm 50\%$ with a nominal value of 20 mA) and a replicate pulse phase margin of 30 degrees (833 nsec for a 100-kHz drive frequency). These margins are sufficient from the point of view of system design requirements; the pulse amplitude must have a margin of 10–15%, and the timing must have a margin of from 100 to 200 nsec. These function pulse margins do not change when the drive field is reduced to 45 Oe or increased to 65 Oe, although maximum bias margins are dependent on the drive fields.

Figure 39 shows dependence of output voltage and noise level on the drive field at chip temperature of 0, 33, and 90°C. The bias field is fixed at the midpoint of the bias field range. Although the output voltage is reduced at high temperatures (because of reductions in resistance and magneto-resistive

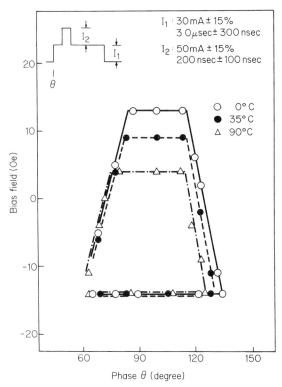

Fig. 38 Overall bias margin versus replicate pulse phase at three chip temperatures. Inset shows pulse shape and amplitude condition (55-Oe peak field of 100-kHz triangular drive).

coefficient), output voltages of more than 6 mV and sufficient signal-to-noise ratios are obtained without regard for the drive field. The resistance of the detector at 33°C is about 1.2 kΩ, and its temperature coefficient is −0.41%/°C. Figure 40 shows temperature dependence of the bias field margins at drive fields of 50 Oe and 65 Oe in a packaged device. The bias field is provided by a pair of ferrite magnets in order to compensate for the temperature dependence of the operating bias field. The field from the magnets has a temperature coefficient of −0.19%/°C and the bias field margin is tracked quite well, as shown in Fig. 40. In this case, the chip is almost fully loaded with 2016 pages in the worst-case data pattern, and 33 loops are masked as bad loops. Although the figure shows only the dependence on package case temperature from −8 to 86°C (−2 to 92°C for chip temperature), in general the bias field margin tends to be limited beyond this range by reduction of the output signal, instability of start–stop

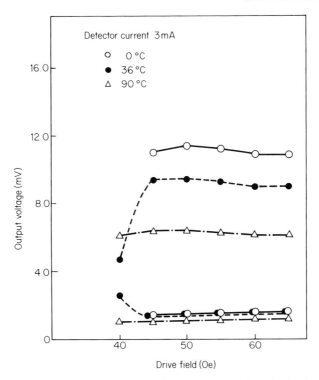

Fig. 39 Output signal and noise voltages versus drive field at three chip temperatures (100-kHz triangular drive).

operation, various kinds of strip-out errors (at high temperatures), and (at low temperatures) upper threshold deterioration in the minor loops. These limiting factors mainly depend on processing and pattern design details.

Figure 41 shows error rate curves of the bias field thresholds for a 55-Oe drive field and 30°C chip temperature for long-term testing. The error rate is defined as the number of read-out errors divided by the number of read-out bits including zeros. A single error in one test cycle corresponds to a 10^{-6} error rate. The error rate curves are shown for three different conditions: 32-page loading, 2016-page loading (both without start–stop operations), and 2016-page loading with 60,000 start–stop operations at each bit position. The data patterns in the 32-page and 2016-page loadings are the worst-case situations, where the rate of "1" is 60%. The difference between the high-bias threshold curves for 32 pages and 2016 pages shows the bubble loading effect: that is, a long-range bubble–bubble interaction effect. The difference between those of with and without start–stop operations indicates deterioration of high-bias thresholds due to start–stop operations in the minor loops. The

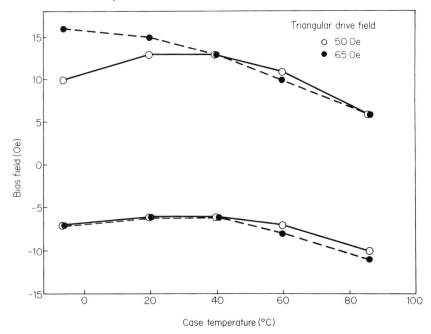

Fig. 40 Temperature dependence of overall bias margins for a packaged device (100-kHz triangular drive).

bias field differences, however, strongly depend on defects in the minor loops and how they are eliminated as bad loops. In the case of Fig. 41, for example, the 32-page curve was measured after eliminating 17 bad loops with an error rate of 10^{-6}, and the 2016-page curve was obtained after eliminating 21 more bad loops with small defects which could not be detected by 32-page loading. Further, the error rate curve for 2016 pages with start–stop operations was measured after eliminating four more bad loops with the defects which could not be detected without start–stop operations. The threshold values of the curves for both 2016-page loadings saturate very well at error rates below 10^{-6}, both for high-bias fields and low-bias ones, and the slopes are less than 0.25 Oe/decade. Linear extrapolation of these slopes indicates that an error rate better than 10^{-14} cm can be guaranteed by the testing for error rate of 10^{-6} if a bias margin of more than 2 Oe is provided for further threshold degradation.

These characteristics of the 1-Mbit chip compare quite well with those of 256-kbit chips with a 16-μm period and 3-μm diameter bubbles. Figure 42 shows examples of overall operating margin curves which directly compare the 1-Mbit chip and the 256-kbit chip under the same conditions (at room temperature and with the 32-page worst-case date pattern). Almost the same

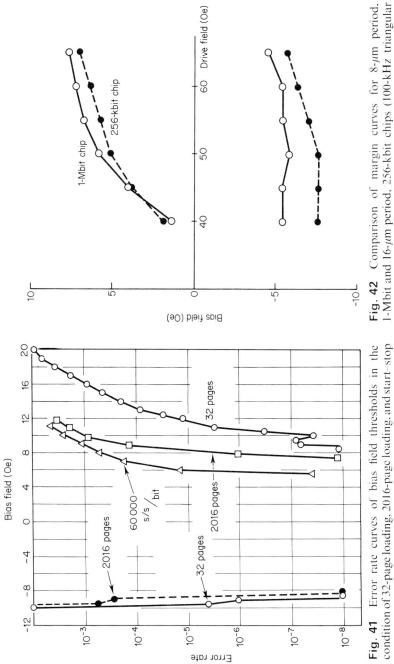

Fig. 41 Error rate curves of bias field thresholds in the condition of 32-page loading, 2016-page loading, and start–stop (S/S) operations (55-Oe, 100-kHz triangular drive, 30 C).

Fig. 42 Comparison of margin curves for 8-μm period, 1-Mbit and 16-μm period, 256-kbit chips (100-kHz triangular drive, 32 pages).

Table 4 Comparison of the 8-μm period 1-Mbit chip with the 16-μm period 256-kbit chip.

Characteristic	Bubble diameter (μm)	
	1.9	3.0
Cell size (μm)	7 × 8	17 × 14
Minimum feature size (μm)	1.0	2.0
Registration tolerance (μm)	±0.5	±0.5
Conductor width (μm)	4.0	4.0
Chip size (mm^2)	9.1 × 9.9	10 × 9.5
Drive field (Oe)	50–65	50–65
Drive frequency (kHz)	100	100
Average access time (msec)	11.2	6.0
Data rate (kbit/sec)	100	100

bias margins (in percentages of nominal bias) are obtained over the same drive field range. The main specifications and performance of the 1-Mbit chip are summarized and compared with those of the 256-kbit chip in Table 4. Bit density quadrupled with the same process technology (except for the 1-μm minimum feature size) and with the same performance (except for access time). These results show that the problems in realizing 8-μm period, 1-Mbit devices have been solved by the design considerations and optimization described thus far (and especially by the concept of relaxed function designs), and that even higher density and capacity could be realized in permalloy bubble memory devices.

7 FURTHER DEVELOPMENT OF PERMALLOY DEVICES

The pursuit of even higher bit density is endless. This section describes the situations of new device technologies intended for realizing high-density devices with the capacity of 4 Mbit or more. In particular, it describes the development of a newly proposed wide-gap propagation pattern and its application to a 4-μm period permalloy device (Yanase *et al.*, 1983).

7.1 New technologies for higher density devices

The results of study of problems and their countermeasures for 8-μm period devices can be applied to the development of higher density permalloy devices, such as 4-μm period, 4-Mbit capacity devices. Although the problems encountered there would be far more severe, the countermeasures discussed would still be useful in solving them. When the design for the 8-μm period

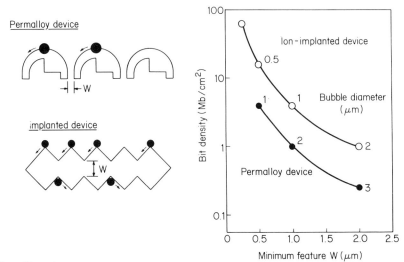

Fig. 43 Bit density versus minimum feature size for permalloy devices and ion-implanted devices (W = width).

device is scaled down for 4-μm period devices, the most difficult problem to overcome is the limitation of lithographic systems. It is very difficult to exceed resolution of 1 μm in photolithography, while new technologies, such as E-beam or x-ray lithography, are not yet ready for use in the production. As shown in Fig. 43, the scaling law predicts a required minimum feature size of 0.5 μm for 4-μm period, 4-Mbit devices, and 0.25 μm for 2-μm period, 16-Mbit devices. On the other hand, the required minimum feature size in ion-implanted devices is significantly relaxed, and a 1-μm size is required for the 4-Mbit devices. Because of this difference in requirements, extensive efforts have been devoted to the development of ion-implanted devices to attain higher densities.

This new technology, however, still faces difficulties in realization of some functions. This is especially true of replication. Such functions could be realized fairly easily with permalloy devices and, based on the technological background accumulated so far, efforts have been made to develop permalloy 4-Mbit devices at the expense of chip size. With the prospect of attaining a minimum feature size of 0.7–0.8 μm through the recent development of projection photolithographic systems, development of 4-Mbit chips with propagation periods of 5–6 μm and with a chip size of about 2 cm^2 has been reported (Fontana *et al.*, 1980; Gergis *et al.*, 1980; George *et al.*, 1981b). Here the serious problem is the increase in the minimum required drive field. The overall bias margin at low drive fields is already limited in 1-Mbit devices with the double-period gate design by minor-loop propagation, and

Field Access Permalloy Devices 195

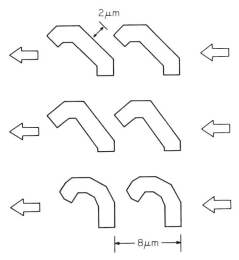

Fig. 44 Bobeck's new propagation pattern designs for an 8-μm period propagation track with 1.9-μm bubbles.

a considerable increase in drive field is inevitable even when submicrometer features are considered and large chips are allowed.

Apart from scaling, some novel structures have been proposed for relaxing the minimum feature size requirement for permalloy devices. The first of these is double-layered propagation patterns, proposed in a rather early stage (Matsuyama *et al.*, 1975). Each of the neighboring propagation patterns is formed in a different layer, and the gap between them is defined by the registration between the layers. In this case, accuracy of pattern delineation and registration, instead of lithographic resolution, would limit propagation characteristics. Another novel structure is called a permalloy contiguous-disk device (Cohen *et al.*, 1979) or complementary permalloy propagation structure (Gergis *et al.*, 1979a). Contiguous permalloy disk patterns without gaps are formed with a small vertical gap to the surrounding permalloy film. Bubbles propagate along the perimeter of the disk pattern in a manner similar to that in ion-implanted devices and cross the cusp between the disks with the help of a magnetic pole in the surrounding permalloy. Although this structure can be obtained only by depositing a permalloy film on a delineated step, the processing for securing a stable vertical gap while holding spacing between the upper permalloy patterns and the bubble film within some appropriate range is difficult. Further, there have been no ideas about realizing functions such as gates and detectors.

Recently, an impressive new pattern design has been proposed by Bobeck *et al.* (1981). Some examples of the pattern design are shown in Fig. 44.

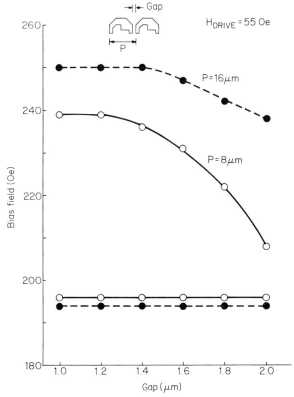

Fig. 45 Gap-size dependence of propagation bias margins in 16-μm and 8-μm half-disk tracks with 1.9-μm bubbles (55-Oe, 100-kHz triangular drive).

They are called "wide-gap" patterns and can propagate 2-μm diameter bubbles at relatively low drive fields, even with 2-μm gaps in an 8-μm period track. The performance of the "wide-gap" pattern indicates a good possibility of realizing 4-μm period permalloy devices with a 1-μm minimum feature size, provided function designs are devised to match the propagation pattern design. One suggestion would be to employ the enlarged function pattern designs used in the 1-Mbit device. Figure 45 shows the gap tolerance of the double-period half-disk pattern. For 2-μm bubble propagation, the 16-μm period half-disk pattern with 2-μm gaps shows a bias margin which is comparable to that of the 8-μm period half disk pattern with 1-μm gaps.

7.2 8-μm period "wide-gap" track and its scaling

First, the large gap tolerance of the "wide-gap" pattern was confirmed and analyzed in the experiments using an 8-μm period track and 1.9-μm bubbles.

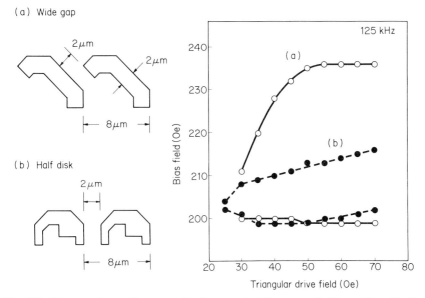

Fig. 46 Propagation performance for 8-μm period, 2-μm gap "wide-gap" and half-disk tracks.

Figure 46 shows the designs of 8-μm period, 2-μm gap "wide-gap" and half-disk tracks, and their straight propagation margins at 125 kHz triangular wave drive. A sufficient bias margin and a low required drive field are obtained in the "wide-gap" track, while a poor bias margin is obtained in the half-disk track. A garget-to-permalloy spacing of 4300 Å and permalloy film thickness of 3700 Å are used in this experiment.

In order to understand such large gap tolerance of the "wide gap" track, real bubble motions were observed in a quasi-static operation and potential well depths were measured. Figure 47 shows bubble collapse fields at each phase of a 58-Oe rotating drive field for the 8-μm period "wide-gap" tracks. The collapse field at point B of the pattern, as shown in the figure, gradually decreases as the drive field is rotated to 337·5°, while the collapse field at point C of the next pattern has already become higher than the free-bubble collapse (FBC) at 315° and gradually increases with further drive field rotation. After the phase point of 337.5°, the collapse field at point C becomes much higher than that at point B, and the collapse field at point B takes on the same value as that of the free bubble, which means that point B becomes magnetically neutral. Consequently, sufficient potential gradient to propagate a bubble across the gap can be supplied by repulsive magnetic poles of A, a neutral of B, and attractive poles of C. The superior gap tolerance of the "wide-gap" track can be explained in Fig. 48, which illustrates potential gradient across the pattern gap.

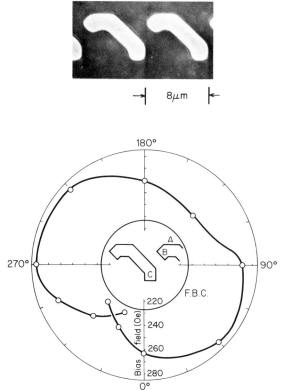

Fig. 47 SEM photograph and potential well depths of the 8-μm period "wide-gap" track.

By comparison, for the conventional half-disk track, deep and wide potential wells around the gap are deformed, presumably by a potential barrier caused by widening the gap, as shown by the dotted line in Fig. 48a. This potential barrier would prevent most bubbles from stretching and propagating across the gap; only large-diameter bubbles—that is, bubbles of a low bias field range—could get across it.

On the other hand, for the "wide-gap" track, the above-mentioned magnetic pole formation is supposed to give a smooth potential gradient when the gap is increased as shown in Fig. 48b. From these studies, it is assumed that a designing point for a 4-μm period "wide-gap" track is to make the potential gradient between the elements as steep and smooth as possible, by utilizing the attractive, neutral, and repulsive magnetic poles.

A 4-μm period "wide-gap" track which is half scaled from the 8-μm period "wide-gap" track shown in Fig. 46 was tested. Figure 49 shows an SEM

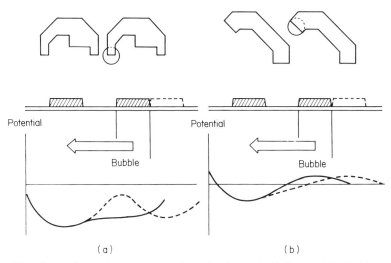

Fig. 48 Illustrations of potential gradient for the (a) half-disk and (b) "wide-gap" tracks.

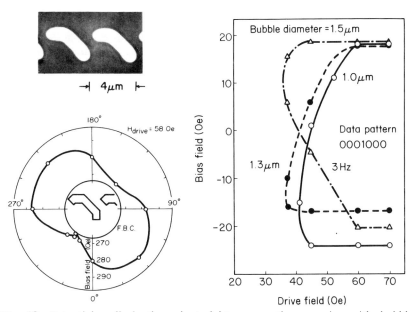

Fig. 49 Potential well depth and straight propagating margins with bubble diameters of 1.0, 1.3, and 1.5 μm for the 4-μm period "wide-gap" track.

photograph of this track and the straight propagation margins in a quasi-static operation with bubble diameters of 1.0, 1.3, and 1.5 μm. The garnet-to-permalloy spacing is 2300 Å and the permalloy film thickness is 3700 Å in this case. Patterning of the permalloy tracks was done by a 10:1 projection step and repeat on wafer system. Although relatively good margins are obtained for isolated bubbles, as shown in Fig. 49, consecutive bubbles do not propagate, regardless of bubble size. To find the reason why the consecutive bubbles do not propagate in this 4-μm period track, its potential well depths were measured and compared to those for the 8-μm period track shown in Fig. 47.

In this 4-μm period track, the collapse fields at both left and bottom ends of the pattern are just barely higher than the free-bubble collapse field at 337.5°, as also shown in Fig. 49. Therefore, sufficient potential gradient to propagate the bubble across the gap can be obtained at a considerably later phase of the drive field than in the 8-μm period track. This is believed to be caused by the deviation of the fabricated shape from the design shape. From these results, it is assumed that in order to propagate consecutive bubbles in the 4-μm period "wide-gap" track, each pattern end should be made as long as possible in the 0° and 315° directions. Following this design rule, scores of 4-μm period "wide-gap" tracks were designed and tested by using 1.3-μm bubbles, which are supposed to be suitable for 4-μm period tracks, judging from the propagation performances shown in Fig. 49.

7.3 Optimization of 4-μm period "wide-gap" track

Figure 50 shows one of the best designs and its straight propagation margins for three consecutive and isolated bubbles at 100-kHz triangular wave drive. The same layer structure as that in the case of Fig. 49 has been employed. As shown in the SEM photograph, the fabricated track retains its shape. Although in this design a sufficient propagation margin for the isolated bubbles can be obtained, the drive field is increased and the upper threshold of the bias margin for the consecutive bubbles is drastically decreased. Microscopic observation shows that the required drive fields for three consecutive bubbles are determined by the middle one pushed out of the track during propagation from the bottom to the top position of the pattern, and the upper threshold is determined by the bubble collapse at the pattern gaps. The "push out" error mode is believed to be caused by the shallow potential along the "leg" of the pattern and to be eliminated by increasing the width of the "leg", which decreases the demagnetizing effect. The collapse mode at the gaps may be caused by insufficient potential gradient between adjacent patterns.

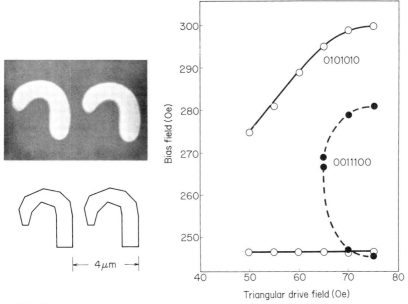

Fig. 50 One of the improved designs for the 4-μm period track and its straight propagation margins for three consecutive and isolated bubbles.

Figure 51 shows three different designs of the 4-μm period track with a wide "leg" and a reduced distance between adjacent patterns, and their straight propagation margins. Compared to the design and characteristics for the 4-μm period track shown in Fig. 50, the required drive fields for all designs can be considerably reduced and their bias margins can be greatly improved even for the consecutive bubbles. Microscopic observation shows that in design BE and BC, lower thresholds of the bias margins at low drive fields are still determined by the "push out" errors, and in all designs, upper thresholds of the bias margins at low drive fields are determined by another error mode in which the last bubble of the consecutive ones is stepped back 1-bit position during propagation across the gaps. Such "stepped back" errors are thought to be caused by insufficient potential gradient at the gaps. In order to eliminate this error, the "arm" length of design BC is varied as shown in Fig. 52. This figure also shows their straight propagation margins. Design BA, with a too-short "arm", is found not only to degrade the upper threshold of the bias margin but also to increase the required drive field. Conversely, design ED, with a too-long "arm", is found to degrade both the upper and lower thresholds of the bias margin, although its required drive field is reduced. This result indicates that the "wide-gap" track with a long "arm" becomes more like "half-disk". Consequently, sufficient bias

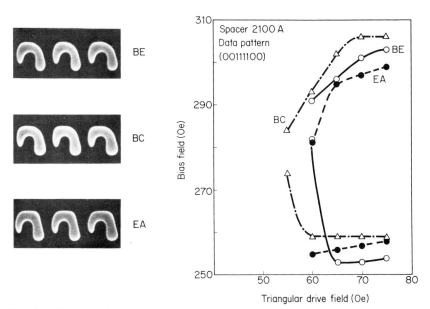

Fig. 51 Designs of the 4-μm period "wide-gap" tracks with a wide leg and their propagation margins.

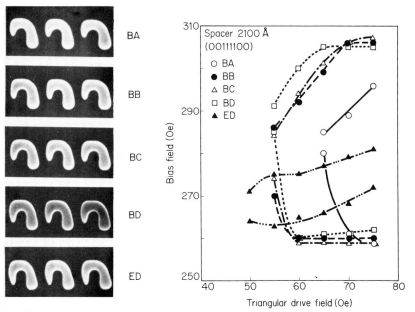

Fig. 52 Designs and propagation margins of the 4-μm period tracks with various arm lengths.

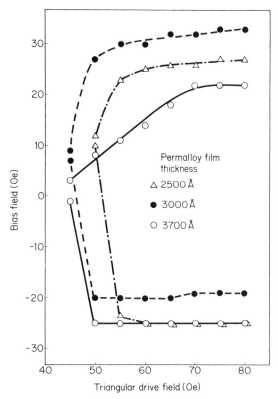

Fig. 53 Permalloy film thickness dependence of the 4-μm period propagation.

margins and low required drive fields can be obtained in the 4-μm period "wide-gap" track with a moderate "arm" length, such as in designs BB, BC, and BD. In these designs, however, both "push out" and "stepped back" errors are still observed at lower drive fields.

In addition to pattern design considerations for the 4-μm period permalloy tracks, thickness dependence of the permalloy film was studied using the 4-μm period track shown in Fig. 50. Garnet-to-permalloy spacing of 2300 Å and 1.3-μm bubbles with an isolated data pattern were used. As shown in Fig. 53, the propagation margin of the 4-μm period track is more sensitive to the permalloy film thickness than expected from the experience in the 8-μm period half-disk device development. In a 3700-Å track, an upper threshold of the bias margin at lower drive fields is observed to be considerably degraded, which would be caused by bubbles idling around the pattern itself. In a 2500-Å track, the required drive field is increased, although the degradation of the upper threshold of bias margin at the lower drive fields is

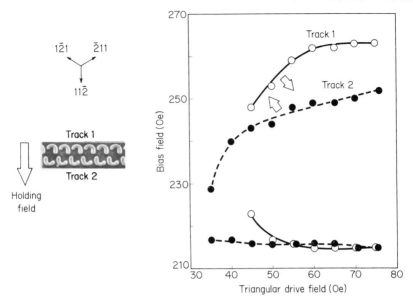

Fig. 54 Margin difference between going and returning tracks which are aligned along a direction perpendicular to (11$\bar{2}$).

almost eliminated. From these results, it is concluded that 3000 Å is the optimum permalloy film thickness for this 4-μm period track, using 1.3-μm bubbles.

It has already been shown that the straight propagation margin of the 8-μm period half-disk track depends on its propagation directions with respect to the garnet crystal axes. In the 4-μm period "wide-gap" tracks, the same behavior is observed. Figure 54 shows the straight propagation margins for the going and returning tracks, which are aligned along a direction perpendicular to (11$\bar{2}$). While a required drive field for track 2 is lower than that for track 1, a bias margin at high drive field for track 1 is better than that for track 2. The garnet wafer used in this study is implanted with 1×10^{14} Ne$^+$/cm^2 at 50 keV. As in the case of half-disk tracks, these margin differences for the minor-loop tracks, track 1 and 2 can be minimized by selecting their appropriate alignments with respect to the garnet crystal axes and a direction of a dc in-plane holding field which is applied for stabilizing start–stop operation. In Fig. 54, a dc in-plane holding field applied in the (11$\bar{2}$) direction degrades the bias margin for the superior track 1 but improves that for inferior track 2, reducing the bias margin difference between the tracks. In the following, the data of straight propagation margins are measured on the tracks which correspond to track 1 in Fig. 54 with holding field of 5 Oe, when not otherwise specified.

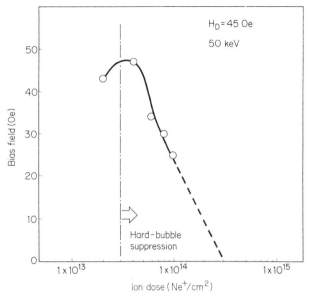

Fig. 55 Propagation margins of the 4-μm period track as a function of Ne ion dose for hard-bubble suppression.

The bias margin of the straight propagation in the 4-μm period "wide-gap" track shown in Fig. 50 was found to be significantly dependent on ion doses for hard-bubble suppression. In Fig. 55 the bias margins of this 4-μm period tracks at a drive field of 45 Oe are shown as a function of Ne^+ dose at implantation energy of 50 keV. The experiments of wide-gap tracks were started with the ion dose of 1×10^{14} Ne^+/cm^2 at 50 keV, which is the ion implantation condition for the 8-μm period half-disk device. When the ion dose is reduced to 3×10^{13} Ne^+/cm^2, a significant improvement in the bias margin is obtained as shown in Fig. 55. This dose corresponds to the threshold for hard-bubble suppression measured by the collapse field dispersion in microscopic observation. On the other hand, the bias margin is degraded very sharply as the ion dose increases. With the ion dose of 3×10^{14}, no bias margin is obtained at any drive field. No bias margin is also observed with the ion dose of 1×10^{14} Ne^+/cm^2 at 70 keV.

It is supposed that the degradation of the bias margins with increasing ion dose is related to the closure domains created in the implanted surface layer of the garnet film.

As already described, the straight propagation margins of the 4-μm period tracks were found to be improved not only by the pattern designs but also by certain other conditions, such as permalloy film thickness, crystal orientation, and ion implantation.

It was also found that a more effective way to reduce the required drive fields and improve the bias margins for the 4-μm period tracks is to decrease garnet-to-permalloy spacing. Figure 56 shows the straight propagation margins of 10 consecutive 1.3-μm bubbles in the design BB track, shown in Fig. 52, for five thicknesses of garnet-to-permalloy spacing. The required drive fields for tracks under 2000 Å are found to be greatly reduced with sufficient bias margins. On the other hand, for tracks of 2000 Å or more, the drive fields are found to be greatly increased as shown in the figure, caused by the "stepped back" errors of the last bubble of 10 consecutive ones. Although it appears to be better to keep the spacing as small as possible, both the upper and lower thresholds of the bias margins for tracks

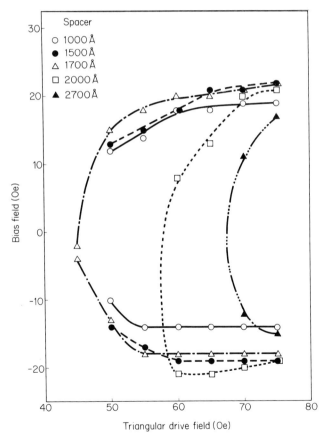

Fig. 56 Spacing thickness dependence of the propagation margin with 10 consecutive bubbles for the 4-μm period track, design BB.

under 1000 Å are apt to be degraded. For the design BB track, the optimum garnet-to-permalloy spacing is found to be between 1500 and 1700 Å.

In summary, good performance for the 4-μm period permalloy track can be obtained by using 1.3-μm bubbles, design BB, permalloy film thickness of 3000 Å, appropriate crystal orientation and direction of dc in-plane holding field, ion implantation for hard-bubble suppression with 5×10^{13} Ne$^+$/cm^2 at 50 keV, and garnet-to-permalloy spacing of 1600 Å.

7.4 Design and characteristics of a 4-μm period major–minor chip

In order to obtain a folded minor-loop organization, 180° turns were designed and tested. Primary designs for two sorts of 180° turns, 1a and 2a shown in the insets of Fig. 57, which are half-scaled from those designed for an 8-μm period device, have high drive fields and poor bias margins for 4-μm period tracks, as shown in Fig. 57. For the 180° turn 1, a sufficient bias margin and a low required drive field are obtained by changing the design into 1(b), as shown in Fig. 57. For the 180° turn 2, the required drive field can be greatly reduced by enlarging the permalloy volume of the corner pattern, as shown in design 2(b).

Using these (b) designs for 180° turns 1 and 2 and the results described

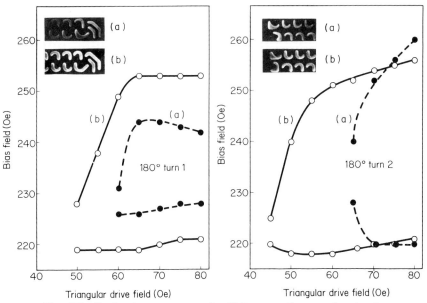

Fig. 57 Design improvements for 180° turns. See text for details.

Fig. 58 SEM photographs of the block-replicate–swap organization chip fabricated with a dual spacer.

in the previous section, a 30-kbit chip with the relaxed function design and the folded minor-loop organization was designed and fabricated. Figure 58 shows SEM photographs of the fabricated 30-kbit chip. Designs of block-replicate gates and true swap gates are almost the same as those of the 1-Mbit device described in Section 6. The basic cell size is $4 \times 4.7\,\mu m$, and the pattern period of write and read major lines is $14\,\mu m$. The minimum

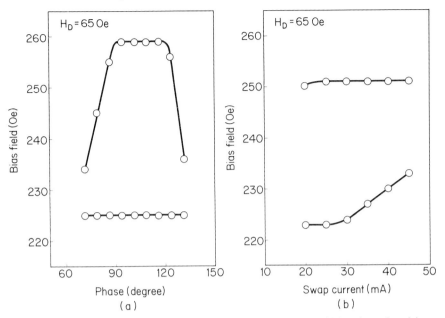

Fig. 59 (a) Bias margin versus replicate pulse phase at 100-kHz triangular drive field of 65 Oe. (b) Bias margin versus swap current at 100-kHz triangular drive field of 65 Oe.

conductor widths and gaps are 4 and 1 μm, respectively. The hammer-type nucleation generator and the 10.5-μm period asymmetric chevron detector are of the same designs as those of the 1-Mbit device.

The composition of the garnet film used in this study is $(YSmLuCa)_3$-$(GeFe)_5O_{12}$, and bubble diameter is 1.3 μm. PLOS is used for planarizing conductor patterns. Here, dual spacing was introduced for the layer structure of the chip. Garnet-to-permalloy spacing is 1600 Å for the 4-μm period minor loops to improve their propagation performance, and 4000 Å for the major lines with the relaxed function design to eliminate spontaneous bubble nucleation at long-period patterns. For the minor loops, 1600-Å spacing is achieved by ion milling after fabrication of 4000-Å spacing for the major lines.

The fabricated 30-kbit chips were characterized with 100-kHz triangular wave drive. The overall bias margin is 20 Oe at the triangular drive field of 65 Oe with 10 consecutive pages. Figure 59a shows the bias margin for the block replicate gates as a function of the replicate pulse phase at a drive field of 65 Oe. The phase margin is 30 degrees, almost the same as that for the 1-Mbit device. Figure 59b shows the bias margin for the true swap gates as a function of the swap current at a drive field of 65 Oe. Current margin

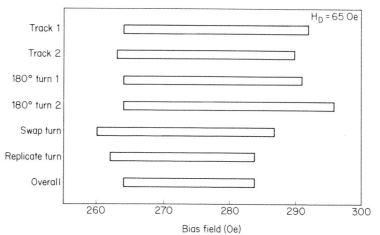

Fig. 60 Propagation characteristics of the folded minor loops and the overall bias margin at 100-kHz triangular drive field of 65 Oe.

is 25 ± 5 mA. Minimum swap current of this device is 10 mA greater than that of the 1-Mbit device, due to the pickax pattern used in the major line instead of a half-disk pattern in the 1-Mbit device.

The propagation margins for closed minor loops determine the overall bias margin of this chip, while the major line propagation and the function operations have good performance. Figure 60 shows results obtained in diagnostic testing for the minor-loop propagation with 10 consecutive pages at a drive field of 65 Oe. The 4-μm period tracks, tracks 1 and 2 (bubbles propagate in the direction of the swap gates and the replicate gates, respectively), optimized as described earlier, have good performances. Further, the propagation margins for 180° turns 1 and 2 (located at the center of the minor loops and near the swap and replicate gates, respectively) are improved by having used the results shown in Fig. 57. The upper thresholds of the propagation margins for both the swap and replicate turns determine the overall bias margin of this chip. These margin-limiting factors are expected to be eliminated by the improvements of pattern design and fabrication process in further development.

From these results, a practical 4-Mbit bubble memory device is expected to be realized by using a 4-μm period permalloy device.

REFERENCES

Almasi, G. S., and Lin, P. S. (1976). *IEEE Trans. Magn.* **MAG-12**, 160–202.
Archer, J. L. (1977). *Dig. Intermag Conf.* No. 11-1.

Bobeck, A. H., and Danylchuk, I. (1977). *IEEE Trans. Magn.* **MAG-13**, 1370–1372.
Bobeck, A. H., and Della Torre, E. (1975). "Magnetic Bubbles." North-Holland Publ., Amsterdam.
Bobeck, A. H., Fisher, R. F., and Smith, J. L. (1972). *AIP Conf. Proc.* No. 5, 45–51.
Bobeck, A. H., Danylchuk, I., Rossol, F. C., and Strauss, W. (1973). *IEEE Trans. Magn.* **MAG-9**, 474–480.
Bobeck, A. H., Chirovsky, L., Shapiro, H. M., and Wagner, R. S. (1981). *Magn. Magn. Mater. Conf., Atlanata, Ga.* Pap. EA-1.
Bonyhard, P. I. (1977). *IEEE Trans. Magn.* **MAG-13**, 1785.
Bonyhard, P. I. (1979). *J. Appl. Phys.* **50**, 2213–2215.
Bonyhard, P. I., and Smith, J. L. (1976). *IEEE Trans. Magn.* **MAG-12**, 614–617.
Bonyhard, P. I., Danylchuk, I., Kish, D. E., and Smith, J. L. (1970). *IEEE Trans. Magn.* **MAG-6**, 447–451.
Bonyhard, P. I., Geusic, J. E., Bobeck, A. H., Chen, Y. S., Michaelis, P. C., and Smith, J. L. (1973). *IEEE Trans. Magn.* **MAG-9**, 433–436.
Bonyhard, P. I., Chen, Y. S., and Smith, J. L. (1974). *AIP Conf. Proc.* No. 18, 100–104.
Bonyhard, P. I., Chen, Y. S., and Smith, J. L. (1977). *IEEE Trans. Magn.* **MAG-13**, 1258–1260.
Bullock, D. C., Fontana, R. E., Singh, S. K., Bush, M., and Stein, R. (1979a). *IEEE Trans. Magn.* **MAG-15**, 1697–1702.
Bullock, D. C., Fontana, R. E., Singh, S. K., Seitchik, J., and Closson, A. (1979b). *J. Appl. Phys.* **50**, 2222–2224.
Cohen, M. S., Kane, S. M., and Sanders, I. L. (1979). *IEEE Trans. Magn.* **MAG-15**, 1654–1656.
Davies, J. E., Clover, R. B., Lieberman, B., and Rose, D. K. (1980). *IEEE Trans. Magn.* **MAG-16**, 1106–1110.
Dimyan, M. Y., and Hubbell, W. C. (1979). *J. Appl. Phys.* **50**, 2225–2227.
Dishman, J. M., Pierce, R. D., and Roman, B. J. (1974). *J. Appl. Phys.* **45**, 4076–4083.
Fontana, R. E., Bullock, D. C., and Singh, S. K. (1980). *IEEE Trans. Magn.* **MAG-16**, 1101–1105.
George, P. K., Hughes, A. J., and Archer, J. L. (1974). *IEEE Trans. Magn.* **MAG-10**, 821–824.
George, P. K., Gill, H. S., Moberly, L., and Norton, R. H. (1981a). *IEEE Trans. Magn.* **MAG-17**, 1442–1451.
George, P. K., Gill, H. S., Norton, R. H., Reyling, G. F., and Tuxford, A. M. (1981b). *Dig. Intermag Conf.* No. 9-1.
Gergis, I. S., and Kobayashi, T. (1978). *IEEE Trans. Magn.* **MAG-14**, 1–4.
Gergis, I. S., Lee, W. P., and Salle, C. D. (1979a). *IEEE Trans. Magn.* **MAG-15**, 1719.
Gergis, I. S., Tocci, L. R., Williams, J. L., and Lee, W. P. (1979b). *J. Appl. Phys.* **50**, 2216–2218.
Gergis, I. S., Tocci, L. R., Jones, A. B., Pulliam, G. R., and Reekstin, J. P. (1980). *ICMB, 4th, Tokyo* Pap. D-6.
Guldi, R. L., Fontana, R. E., and Bullock, D. C. (1981). *J. Electrochem. Soc.* **128**, 675–678.
Hiroshima, M., Asana, A., Yoshızawa, S., Saito, N., and Kasai, M. (1976). *Proc. Conf. Solid State Devices* **7**, 113–116.

Igarashi, S., Igarashi, K., Hirano, A., Orihara, S., and Yamagishi, K. (1976). *AIP Conf. Proc.* No. 29, 48–50.
Inoue, H., Asama, K., Komenou, K., and Kashiro, K. (1979). *Proc. Conf. Solid State Devices* **10**, 225–229.
Johnson, W. A., Hagedorn, F. B., Wolfe, R., Vella-Coleiro, G. P., Smith, J. L., and Wagner, R. S. (1981). *J. Electrochem. Soc.* **128**, 1808–1814.
Kita, Y., Yamaguchi, N., Sugie, M., and Yoshizawa, S. (1980). *IEEE Trans. Comput.* **C-29**, 89–96.
Kohara, H., Takahashi, K., Suga, S., and Fujiwara, S. (1981). *IEEE Trans. Magn.* **MAG-17**, 3038–3040.
Kryder, M. H., Ahn, K. Y., Almasi, G. S., Keefe, G. E., and Powers, J. V. (1974). *IEEE Trans. Magn.* **MAG-10**, 825–827.
Lin, Y. S., Dove, D. B., Schwarzl, S., and Shir, C. C. (1978). *IEEE Trans. Magn.* **MAG-14**, 494–499.
Majima, T., Hirano, A., and Orihara, S. (1981). *J. Appl. Phys.* **52**, 2395–2397.
Matsuyama, S., Kinoshita, R., and Segawa, M. (1975). *AIP Conf. Proc.* No. 24, 645–646.
Matsuyama, S., Igarashi, K., Majima, T., and Orihara, S. (1977). *Proc. Conf. Solid State Devices* **8**, 351–354.
Matsuyama, S., Kinoshita, R., Yanase, T., Otake, I., and Imamura, K. (1979). *IEEE Trans. Magn.* **MAG-15**, 1715–1717.
Naden, R. A., Keenan, W. R., and Lee, D. M. (1976). *IEEE Trans. Magn.* **MAG-12**, 685–687.
Nelson, T. J. (1977). *IEEE Trans. Magn.* **MAG-13**, 1773–1785.
Nelson, T. J., Chen, Y. S., and Geusic, J. E. (1973). *IEEE Trans. Magn.* **MAG-9**, 289–293.
Orihara, S., Kinoshita, R., Yanase, T., Segawa, M., Matsuyama, S., and Yamagishi, K. (1978). *J. Appl. Phys.* **49**, 1930–1932.
Orihara, S., Yanase, T., and Majima, T. (1979). *IEEE Trans. Magn.* **MAG-15**, 1692–1696.
Reekstin, J. P., and Kowalchuk, R. (1973). *IEEE Trans. Magn.* **MAG-9**, 485–488.
Roman, B. J., Nelson, T. J., and Smith, J. L. (1979). *IEEE Trans. Magn.* **MAG-15**, 1719.
Sakai, S., Matsuyama, S., and Segawa, M. (1976). *AIP Conf. Proc.* No. 29, 30–31.
Strauss, W., Bobeck, A. H., and Ciak, F. J. (1972). *AIP Conf. Proc.* No. 10, 202–206.
Sugita, Y., Nishida, H., Umezaki, H., Tsumita, N., and Yamada, T. (1979). *ICMB 3rd, Indian Wells*, **00** Pap. 6-1.
Suzuki, R., and Humphrey, F. B. (1980). *IEEE Trans. Magn.* **MAG-16**, 1389–1395.
Takeda, S., Nakajima, M., Kitakoji, T., Okuyama, H., and Murakawa, K. (1980). *Ext. Abstr., Electrochem. Soc. Meet., 158th* Pap. 80-2.
Uchishiba, H., Obokata, T., and Asama, K. (1977). *J. Appl. Phys.* **48**, 2604–2607.
West, F. G., Hubbell, W. C., and Singh, S. K. (1976). *AIP Conf. Proc.* No. 29, 28–29.
Yamaguchi, N., and Hibi, S. (1980). *Int. Reliab. Phys. Symp., Las Vegas, Nev.* Pap. 3.3-1.
Yanase, T., Igarashi, K., and Segawa, M. (1978). *Fujitsu Sci. Tech. J.* **14**, 61–84.
Yanase, T., Majima, T., and Orihara, S. (1979). *J. Appl. Phys.* **50**, 2234–2236.
Yanase, T., Inoue, H., Iwasa, S., and Orihara, S. (1980). *IEEE Trans. Magn.* **MAG-16**, 855–857.

Yanase, T., Inoue, H., Iwasa, S., Orihara, S., and Matsuyama, S. (1981). *J. Appl. Phys.* **52**, 2398–2400.

Yanase, T., Amatsu, M., Inoue, H., and Orihara, S. (1983). *Dig. Intermag Conf.* No. FA-1.

Yoshimi, K., Yoshioka, N., Urai, H., Morimoto, A., and Wada, Y. (1978). *J. Appl. Phys.* **49**, 1918–1923.

Ypma, J. E., Gergis, I. S., and Archer, J. L. (1976). *AIP Conf. Proc.* No. 29, 51–53.

5 Circuit Design and Properties of Patterned Ion-Implanted Layers for Field Access Bubble Devices

T. J. NELSON

Bell Communications Research
Murray Hill, New Jersey

D. J. MUEHLNER

AT & T Bell Laboratories
Murray Hill, New Jersey

1	Introduction	216
	1.1 Discovery of contiguous-disk bubble circuits	216
	1.2 Advantages of contiguous-disk bubble circuits	219
	1.3 Structure of contiguous-disk bubble devices	221
2	Chip organization	224
	2.1 G-loop with electronic feedback	227
	2.2 G-loop with nondestructive read out	228
	2.3 Simple major loop	231
3	Active components—transfer	234
	3.1 Field effect transfer	234
	3.2 Gradient effect transfer	235
	3.3 Trapping transfer	239
	3.4 Reverse rotation transfer	242
4	Active components—generation, replication, and detection	245
	4.1 Generation	245
	4.2 Replication	249
	4.3 Detection	251
5	Propagation	256
	5.1 Track orientation	256
	5.2 Track design	259
6	Charged walls	263

7	Damage and strain profiles in the implanted layer	.	.	.	.	271
8	Magnetic properties of the implanted layer	.	.	.	.	277
	References				.	288

1 INTRODUCTION

Shallow ion implantation of magnetic garnets creates a capping layer of reduced uniaxial anisotropy that is useful for the suppression of hard bubbles. If the anisotropy changes sign, and the magnetization lies in the plane in the affected layer, then this magnetization responds to an in-plane drive field much as the magnetization of the permalloy does in permalloy-driven bubble circuits. Bubbles under the implanted layer exhibit an affinity for boundaries with unimplanted regions, and can be made to shift along a sinuous boundary in response to a rotating in-plane field. An advantage ion-implanted circuits have over permalloy ones is that the basic ion-implanted propagation pattern (I2P2) is gapless. The gaps in the permalloy circuits increase the lithographic resolution required and thereby lower the achievable bit density. Moreover, ion-implanted circuits operate at lower drive field than permalloy circuits, so packaged memories can be driven with less power dissipation and temperature rise. Initially workers thought that the implant effect would be of limited use in small-bubble films, so they prepared double films with separate storage and drive layers. By using multiple implantations to change the anisotropy more uniformly throughout the depth of the implanted layer, however, it has been possible to use single-layer films to the highest densities yet attempted. A sensible processing sequence that retains the inherent planarity of the propagation patterns and provides for interconnection between the permalloy sensor and the conductor levels has been developed.

1.1 Discovery of contiguous-disk bubble circuits

Searching for new ways to prepare bubble materials, Wolfe *et al.* (1971) used ion implantation to expand the lattice of rare-earth iron garnets. The implanted layer, which is constrained by the substrate, undergoes expansion only perpendicular to the free surface. This implantation-induced strain contributes in turn to the uniaxial magnetic anisotropy:

$$\delta K_u = \tfrac{3}{2}\lambda_{111}\sigma, \tag{1}$$

where σ is the uniaxial stress. The magnetostriction coefficient λ_{111} is equal to the strain that develops when a randomly magnetized sample is saturated in a [111] direction in a cubic crystal. In a material that has little or negative uniaxial anisotropy ($K_u \lesssim 0$), but positive magnetostriction, ($\lambda_{111} > 0$), the required positive K_u can result from the positive ($\sigma > 0$) stress induced by implantation. Thus a film in which the magnetization is "lying down" might be converted to "standing up" as required for bubble domain stability. However, the primary use of implantation in garnets has been in materials with positive anisotropy and negative magnetostriction.

The discovery of hard bubbles (Tabor et al., 1972) in magnetic garnet films caused investigators to try a second garnet layer that provides a domain-wall "cap" to the base of bubble domains (Bobeck et al., 1972). This cap made the twisted spin arrangement present in the wall of a hard bubble physically implausible. A similar effect was achieved by Wolfe and North (1972) with proton implantation part way into a bubble film ($K_u > 0$) with $\lambda_{111} < 0$. The effect on the bubbles of a dose of 2×10^{16} H ions/cm² at an energy of 100 keV (100/H/2E16, for short) is shown in Fig. 1. Hard bubbles were created in this $(YGdTm)_3(FeGa)_5O_{12}$ film by rapid demagnetization. As can be seen, hard-bubble formation was suppressed in the implanted portion of the sample. The bubbles in the implanted side collapsed at a single critical bias field, but this was not so in the unimplanted side. Some of the domains in the unimplanted region were so hard that they were not even circular. Note that bubbles in the implanted region were attached to the boundary. Although the forces acting on a domain crossing for implant boundary have not been completely analyzed, there is a simple explanation for the tendency of bubbles to stick to the boundary. The magnetization discontinuity (from perpendicular to parallel to the plane of the film) causes a field near the boundary. This field can be thought of as being produced by an Amperian current of $cM\,\delta h$, in Gaussian units (c is the speed of light, M is the magnetization, and δh is the thickness of the step) flowing along the boundary, and acting to attract bubbles from the implanted side and to repel bubbles on the unimplanted side.

At first, it seemed that the edge affinity would be useful to stabilize the positions of bubbles being acted on by various propagation forces (Wolfe et al., 1973). However, it was also apparent that the in-plane magnetization should respond to the drive field, much as in permalloy, and so bubble propagation might result on an appropriately designed track. Naturally, patterns then being designed for permalloy circuits were tried out first. One of them, a line of contiguous circles, had been investigated by A. A. Thiele, who had realized that the gaps of the permalloy patterns represent an important limit. Although the gapless track proved to be a poor permalloy circuit design, it stood out as the best implanted circuit design. Figure 2 shows a slightly more advanced version that Wolfe et al. found propagates bubbles clockwise on the outside, counter-clockwise on the inside, in a clockwise rotating in-plane field. The unimplanted circuit features here were

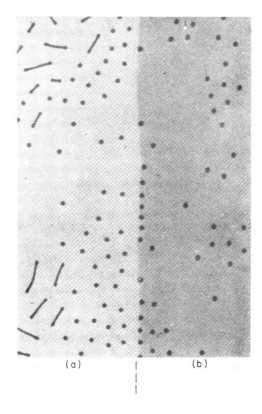

Fig. 1 Photograph showing effects of hydrogen implantation on bubbles in a magnetic garnet film of composition $(YGdTm)_3(FeGa)_5O_{12}$. (a) As grown. (b) Ion-implanted. (After Wolfe and North, 1972.)

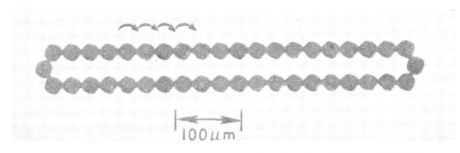

Fig. 2 Ion-implanted bubble propagation circuit. (After Wolfe et al., 1973.)

made visible by the photoresist, which was blackened by the heavy ion implantation.

1.2 Advantages of contiguous-disk bubble circuits

A principal advantage of the ion-implanted circuits is the relative coarseness of the lithography required to form them. Lin *et al.* (1979b) compared minimum feature size w as functions of cell dimension for different versions of bubble circuit technology. Their analysis is summarized in Fig. 3, and shows a clear advantage for ion-implanted bubble circuit technology. At present, permalloy circuits are in production at 64 w^2/cell, while experimental permalloy circuits exist at 20.9 w^2/cell (Bobeck, 1983). Lin *et al.* estimated the resolution requirement in contiguous-disk circuits to be as low as 7.5 w^2/bit. To justify this conclusion, we note designers do not use sharp cusps, as the name "contiguous disk" implies, but often round them out instead. Then the track consists of alternating bulges and hollows with the hollows having the smaller diameter.

With a minimum feature w of $\frac{3}{8}$ of a period λ, as can be seen in Fig. 4, reliable propagation was shown at 8-μm period (Nelson *et al.*, 1979). Note that the cell area is 7.1 w^2/bit, in good agreement with the estimate given above. Half-micrometer bubble propagation was recently been reported at 2-μm period with 12% bias margin range at 50 Oe drive (Sugita *et al.*, 1981). The resolution used to define the 2-μm period track was said to be 0.7 μm, also in agreement with our expectation.

Ion-implanted circuits have additional processing advantages relative to permalloy circuits. Since the implanted circuit has to be patterned first, its lithography is done on a planar substrate, improving resolution. Moreover, the implant masking patterns can be made of photoresist, avoiding the loss of resolution caused by reflections from a metal surface, such as permalloy. The advantage of planarity is not lost for the conductor level, either, because the intervening patterned level, which is the permalloy sensor, is thin.

In field-access bubble devices, power dissipated by the drive coils causes chip temperature to increase, and this narrows the useful temperature range of operation. Of course the coil power increases as the square of the drive field. Kryder (1979) estimated that the minimum drive field for permalloy devices obeys

$$H_{xy} = 0.06(4 + M) + 10 \, \text{Oe}. \tag{2}$$

For ion-implanted devices, the corresponding relationship was argued to be (Calhoun and Sanders, 1981)

$$H_{xy} = 0.37 \frac{t}{\lambda}(4\pi M) + 12.2 \, \text{Oe}, \tag{3}$$

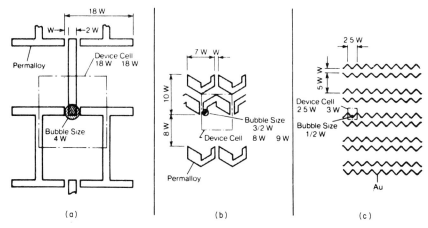

Fig. 3 Areal density comparison between (a) T-bar devices, (b) permalloy asymmetric chevrons, and (c) ion-implanted contiguous disks. (After Lin *et al.*, 1979b.)

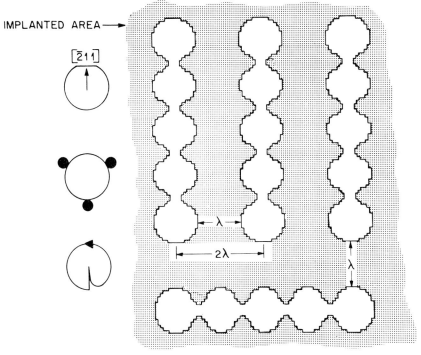

Fig. 4 Unimplanted minor loops on 2λ centers and orientation with respect to crystal axes, preferred positions, and rotating field conventions. (After Nelson *et al.*, 1979.)

where t is the thickness of the implanted layer and λ is the circuit period. The first term comes from the model of Best (1981). With detailed arguments based on spontaneous nucleation, Kryder (1979, 1981) arrived at an estimate for the saturation induction, $4\pi M$, required in either type of circuit versus bubble size: about 625 G for 1-μm bubbles, increasing to 1000 G for 0.5-μm bubbles. Permalloy circuits may not be practical for periods much less than 6 μm. However, Komenou et al. (1981) achieved a 28-Oe bias range with a 4-μm period ion-implanted circuit at drive fields of 45–55 Oe. With Komenou's $4\pi M = 700$ G, and estimating the implant depth at 0.4 μm, the formula predicts a minimum drive field of 38 Oe, which seems reasonably close. On the other hand, the 2-μm period ion-implanted circuit propagation reported by Sugita et al. (1981) had 12% margins at only 50 Oe drive, while the model predicts about 75 Oe. Thus ion-implanted devices have the advantage of moderate minimum drive fields up to the highest densities.

1.3 Structure of contiguous-disk bubble devices

In designing a small bubble material one first finds the material length parameter l and then, based on the required Q (defined as $H_K/4\pi M$) and expected exchange constant A, the anisotropy and saturation induction are determined. The design equations for magnetic bubble materials, as for instance given by Blank et al. (1979), are

$$l = \frac{sw - 0.283h}{6.589 - 0.222(sw/h)}, \qquad (4)$$

where sw is the demagnetized strip width in a film of thickness h and

$$K_u = \frac{4Q^2 A}{l^2} \qquad (5)$$

$$4\pi M_s = \left(\frac{32\pi AQ}{l^2}\right)^{1/2}. \qquad (6)$$

Scaling down with constant A and Q means that the saturation induction and anisotropy increase rapidly. This led Lin et al. (1977) to conclude that the change in anisotropy in implanted small bubble films would be insufficient. To avoid the dilemma they used a second garnet layer, of lower Q, to receive the implant (see Fig. 5). In an alternate approach, Jouve et al. (1976) proposed that a multiplicity of implantations be done with different energies (see Fig. 6). The same total dose could be achieved in this way without

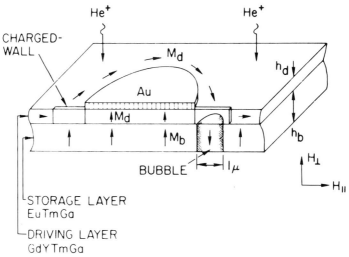

Fig. 5 Schematic drawing of a double-garnet composite, showing the basic propagation mechanism. (After Lin *et al.*, 1977.)

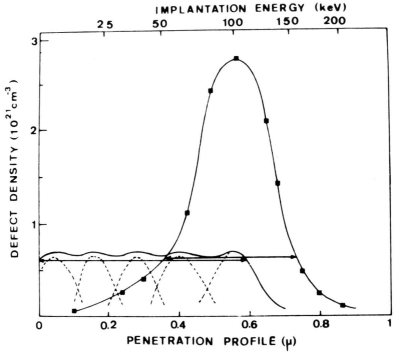

Fig. 6 Experimental defect concentration profile and calculated multiple energy implantation profile, totalling the same dose. (After Jouve *et al.*, 1976.)

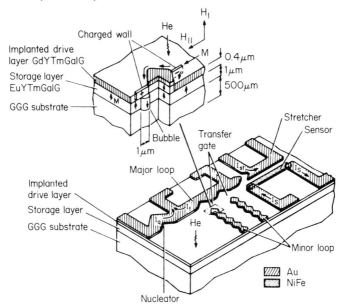

Fig. 7 Schematic perspective view of a self-aligned chip (After Lin et al., 1979b.)

overdosing the material anywhere through the damaged layer. By implanting the same films at successively higher energies, 40, 100, 140, 200/H/1.5E16, Jouve et al. found bubble propagation with wide bias versus drive margin ranges in 2-μm Sm-YIG films for implant depths of 0.5 μm or more. The anisotropy change in small bubble-films also may be increased by suppression of the growth-induced anisotropy at low implant dose (Vella-Coleiro et al., 1981). Therefore, not all of the changes in anisotropy need be produced by magnetostriction.

In some high-density circuits, the gold implant masking material has been made to double as the control circuit (Lin et al., 1979b). The advantage was more than simply to save a few process steps—it guaranteed precise alignment between these two critical levels. Self-aligned processing also eliminates adverse stress effects introduced by a separate conductor crossing the propagate paths. Figure 7 shows designs for a complete set of functions made with self-aligned processing and 1-μm bubbles and operated at 150 kHz. For the time being, however, because of the limited success with circuit design in face of the constraints imposed by the single mask level approach, ion-implanted circuit technology is pursued with separate masking steps for the implant level, a thin permalloy sensor, and a conductor control circuit level. Component design and layout are considerably more flexible using independent patterns in the implant and conductor levels. Figure 8 shows a

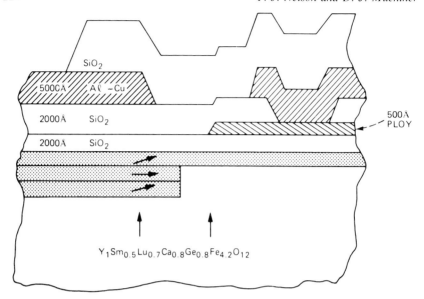

Fig. 8 Schematic cross-section of oxide spacing, permalloy (PLOY), and Al–Cu levels in 8-μm period I2P2 devices. (After Nelson et al., 1980.)

typical device in cross-section (Nelson et al., 1980). Note that the implant consists of a patterned part, as expected, as well as a uniform part similar to that used for hard-bubble suppression. The uniform implant may control wall states when bubble manipulation in the otherwise unimplanted region is done, as in the detector. Additionally, however, the uniform implant was useful in controlling generator currents, about which more will be said below.

Permalloy gates have also been combined with ion-implanted minor loops (Suzuki et al., 1982; Satoh et al., 1983). This was done to combine the block swap/replicate organization, which is difficult to design with ion-implanted circuits, with the high bit density of ion-implanted minor loops. These hybrid circuits are reasonable to attempt mainly because the large permalloy gate structures that have to be used can be accommodated by folding the minor loops.

2 CHIP ORGANIZATION

The block swap/replicate (BS/R) chip organization has become standard in bubble memories based on permalloy circuits. Because integrated controller

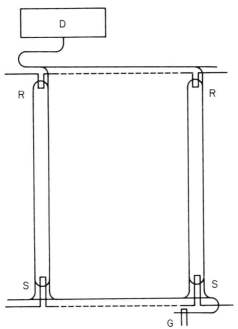

Fig. 9 Block swap/replicate chip organization for bubble memory devices. (After Bonyhard, 1979.)

chips are costly to implement, ion-implanted circuits would be more easily accepted if they would use the BS/R organization also. However, the swap gates (S in Fig. 9) are difficult to design and the replicate gates (R) may be impossible. Therefore we have to consider alternative organizations based on more feasible components, design hybrid circuits, or give up the other advantages of ion-implanted circuits. We think one of the alternative organizations, based on bidirectional transfer gates and called the "shuttle", has no real disadvantages other than being different from the BS/R organization and will become standard in ion-implanted circuits.

One way to retain the BS/R organization, which is being pursued (Suzuki et al., 1982; Satoh et al., 1983), combines ion-implanted minor loops with permalloy swap and replicate gates. Provided the minor loops can be folded, which adds a requirement for a difficult turn design, enough area can be provided for the permalloy gates. But transitions between the permalloy and ion-implanted circuits are needed also. Ikeda et al. (1983) have optimized their implant conditions for bias field compatibility with permalloy functions, while Satoh et al. have cleverly combined the transition into their replicate gate.

Some of the characteristics of the BS/R organization have to be mentioned here so that comparisons can be made in the following discussions of alternative organizations. The essential control components are the generator, the swap and replicate gates, and the detector. The replicate gates, when pulsed, copy a bubble from each minor loop onto a track leading to the detector. This is physically accomplished by passively stretching a bubble along a large permalloy element, actively cutting the strip with the field of a pulsed conductor, and merging one of the resulting bubbles into the read line. Each swap gate, when pulsed, can simultaneously remove an old bubble (if present) from a minor loop and replace it with a recently generated bubble (or vacancy) from the write line. The old bubble then propagates along the write line and is eventually discarded. Because a bubble never leaves its minor loop except to be replaced with a new bubble or vacancy representing updated information, power failures do not disturb the stored information (provided the drive field turns off correctly) even when power fails during read and write operations. The read and write lines could be operated simultaneously, although most user systems probably cannot handle simultaneous read and write data transfers. The BS/R organization places no heavy burden on chip yield, either. For example, in a typical chip with about $m/4$ loops of m steps spaced two periods apart, the length of the read and write lines taken together would be equal to the length of a minor loop. Although the read and write lines must be good for the chip to function at all, this does not limit the yield in a significant way. We take it for granted that redundant minor loops will be used. For purposes of illustration, we will assume the minor loop yield need not be higher than, say, 90%.

We are led by the absence of swap and replicate gates to consider ion-implanted chip organizations in which bubbles are transferred out of the minor loops and must later be transferred back in. Volatility is an important issue, because a power failure must not prevent the eventual replacement of those bubbles. In some of the chip organizations we consider below, bubbles are transferred in and out at the two opposite ends of their minor loops. If the major-line propagate tracks both face the minor-loop ends across the intervening implanted material, then the two tracks propagate bubbles in opposite directions. Lin and Sanders (1981) considered the consequences of data reversal, such as occurred in their self-aligned design, and suggested that it can be handled with error-correcting electronics but did not show their solution explicitly. To avoid having the bit order in blocks as they are read opposite to that in blocks as they are generated, we need to have one major line track facing away from the minor loops. Complementary transfer gates suitable for this organization have been used by Nelson et al. (1980), Komenou et al. (1981), Sanders et al. (1982), and Urai and Yoshimi (1982). Much of the discussion in this section can be found in the paper on chip organization by Bonyhard and Nelson (1982).

2.1 G-loop with electronic feedback

The component layout shown in Fig. 10 closes the path from the detector back to the write line electronically, with the help of the generator. The major loop, if viewed in a particular orientation, resembles the letter "G", hence its name. Complementary sets of transfer gates at both ends of the minor loops connect to major tracks running bubbles in the same direction (left to right in Fig. 10). Other components needed are a generator and a destructive read out (DRO) detector. In operation, a block being read out on the read line simultaneously appears on the write line and is available for replacement soon after the last bit has been detected. The read and write line counts must be equal to the count in the minor loop from the transfer-out gate to the transfer-in gate, so that after the read is completed the bubbles are in position to be returned to the minor loops. Note that the electronic feedback path returns the bubbles much faster than would be possible by propagation. If a power failure should occur during a read cycle, the data leaving the read line and entering the write line should be stopped on predetermined odd or even cycles with respect to the detector. This would guarantee that the process of detection and regeneration, which is likely to be restricted to alternate cycles, could be correctly restarted when power returns. When power is restored, the system controller could complete the reading and rewriting of the block and then circulate it once around the solid path to provide a complete read. To maintain synchronization, the path connecting the end of the write line to the beginning of the read line should have the same count as a minor loop.

The first bit of the block would be identified by a guaranteed bubble,

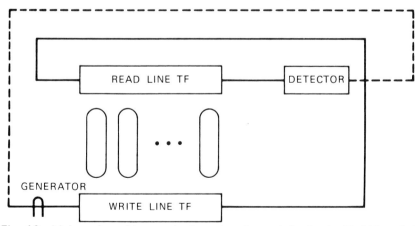

Fig. 10 Major–minor chip organization using electronic feedback of bubble-coded information. (After Nelson *et al.*, 1980.)

which means that we must dedicate one minor loop for synchronization. Because the on-chip feed forward path would normally not be used, and therefore would be empty when the first power failure occurs, the system controller may use the long string of guaranteed zeros to prepare to recognize the first bubble as the block identifier. Repeated power outages might leave the stranded data block in various positions around the G-loop, but it would always be in synchronism with the vacancies in the minor loops and always contain the gap and first loop identifier. However, if a power outage should occur during a write cycle, then the block eventually restored would be partly new and partly old. Some users may find this unacceptable.

The read and write data rates are the same for the BS/R and G-loop organizations, except that read and write operations cannot be carried out simultaneously in the latter. The parts of the G-loop chip that must be good are longer by one minor loop length than in the BS/R chip, because of the path from the write line around to the read line. Thus we would expect that the burden yield of the chip, by which we mean the yield of the parts of the chip that must be good in addition to enough minor loops, could be lowered to about 80%, based on the assumption of a 90% minor loop yield. Also, more directions of propagation are demanded by the path from the write line to the read line. In ion-implanted circuit design, this raises nontrivial (although solvable) problems. For some applications, though, off-chip feedback through the low level output channel provides too weak a form of nonvolatility. We may assume some error correction will be done, so short bursts of infrequent errors are really no problem. However, the stored information would simply be lost if an uncorrectable error (e.g., a long burst) should occur. To avoid this problem, nondestructive readout **NDRO** detection has been proposed.

2.2 G-loop with nondestructive read out

Figure 11 shows a slightly more complicated but similar chip organization in which the feedback path is now completely on chip. This requires a nondestructive detector and a merge gate that permits bubbles from the read line to enter the closed major loop. The merge function has been accomplished in ion-implanted circuits. However, a gap in the unimplanted circuit was used, and a minimum feature of $\frac{1}{4}$ of the circuit period was required. The chip operation is different because we now envision every block of bubbles that is read propagating along the outer path back to the write line. Counting the read line, the overhead is at least 2.5 times the minor loop count m, which we estimate could hypothetically lower the burden yield to 77%. The user data rate for reading many blocks in a row, or writing without reading,

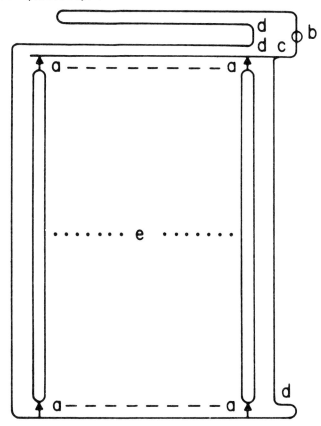

Fig. 11 Circumferential major loop organization. The circuit components required are (a) complementary transfer gates, (b) NDRO detector, (c) merge, and (d) inside turns (e) large storage area. (After Bonyhard *et al.*, 1982.)

is no different from those in the BS/R organization. When changing from a read cycle to a write cycle, or vice versa, the system must wait for blocks to clear the detector and generator and nearby portions of the circumferential major loop. This lowers the effective user's data rate by an amount we have not estimated but do not expect to be significant.

We assume that more than one block and at times up to three blocks will be present on the feedback path. However, there is no way to predict whether the bits of a given block have to be detected on even or odd cycles after a power failure. Consequently, the detector must preserve data that arrive between detect cycles. The generator is to be positioned so that a block may be completely written before the old bubbles are transferred out onto the read line. Since the generator is on the major line, a partially generated

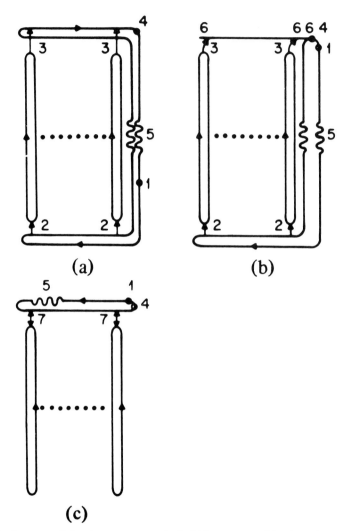

Fig. 12 Layout of circuit components in bubble memory chips based on transfer and NDRO. Separate transfers that retain the data order are shown in (a) and (b), while (c) uses bidirectional transfer. 1, Generator; 2, transfer-in; 3, transfer-out; 4, NDRO detector and annihilator; 5, extra steps (to adjust count); 6, merge; 7, bidirectional transfer. (After Bonyhard and Nelson, 1982.)

block present after power restoration would have to be discarded. To recognize that a block is incomplete, guaranteed bubbles at the trailing edge as well as the leading edge of normal blocks could be provided.

The gap created by the normally unused path from the write line to the read line is needed for the controller to be prepared to recognize the head of a block. Now the distance from the write line to the read line must be large enough to contain two blocks (one from the read line plus a space bigger than a block). This is likely to be greater than $0.5\,m$, which would neatly cover the distance at the minor loop period. The next available choice is $1.5\,m$, since the count from the read line back to the write line must be the same in the major loop as in the minor loops and because the total major-loop count must be an integral multiple of the minor-loop count. If the count from transfer-out to transfer-in can be made as small as $0.5\,m$, the major-loop count will be $2\,m$. However, it may be necessary to let the major loop count be three times the major-loop count with a further reduction in the burden yield. When recovering from a power failure, data present on the major loop must be read to be correctly restored. It is possible here to misidentify a block because of a detection error, and the ensuing incorrect transfer could cause several blocks to be spoiled. This reappearance of weak nonvolatility can be eliminated by reading the data several times before taking any action based on it. Complementary transfer gates that carry bubbles across the major loop after transfer-out as suggested in Fig. 12a and b were introduced by Nelson *et al.* (1981). Bonyhard *et al.*, (1980) discussed operation of 0.5-Mbit chips using these layouts.

2.3 Simple major loop

Bidirectional transfer automatically avoids the problem of data reversal. Figure 12c shows the arrangement of components discussed by Bonyhard and Nelson (1982). It uses a "trapping" transfer design that requires low current and also has low resistance per gate. Provided the chip is square—i.e., it has about $m/4$ loops of length m—the simple major loop SML has the same number of steps and so has the same burden yield as the BS/R organization of the same number and size minor loops.

There is a performance penalty though, as the SML requires restore after read and removal of old blocks during write. To read about $m/4$ user bits, already lined up in the transfer gates, one must propagate for m cycles, which is about a factor of two worse than the BS/R case. Moreover, there is a volatility problem, as will be shown with the help of Fig. 13. On transfer-out, the bubble moves from minor-loop position 0 to major-loop position 1 in one cycle. During transfer back in, the bubble starts in the position

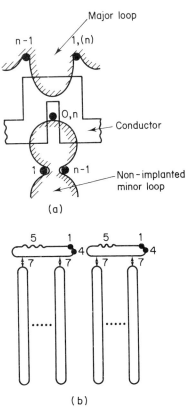

Fig. 13 (a) Count considerations at a bidirectional transfer gate. (b) Split chip shuttle. (After Bonyhard and Nelson, 1982.)

labeled $n-1$ and ends in position n, which is equivalent to 0. From the labeling, one can see that the major loop must be $n-1$ steps long, one cycle shorter than the minor loop. The reason for the difference is that a bubble spends half a cycle being trapped during transfer each way, which adds an effective extra cycle in the major loop. Bubbles have to be transferred back into the minor loops on the first opportunity, as thereafter they would no longer line up with the vacancies they left in the minor loops. However, data left on the major loop after a power failure must be completely read for block identification before it can be restored, and then it is too late. Even if we could correctly identify the block, we could only do it once, which opens the possibility of a detection error leading to misidentification. Thus, at best the SML organization discussed up to this point is not entirely nonvolatile.

There is a way out of this volatility dilemma, however. If time were reversed, we could imagine watching the bubble retrace its steps from 1 on the major loop back to 0 in the minor loop. Provided the major loop count were equal to an integral multiple of the minor loop count, the bubble and vacancy would return to this configuration periodically. Physically we can get the same effect by rotating the drive field backward during transfer-in. Once bidirectional propagation is allowed, secure write becomes possible too. This operation begins with the generation of the new block. The new bubble starts from position 1 on the major loop when the old bubble is at position 0 in the minor loop. If it were not for the old bubble being in the way, the bubble on the major loop could have been transferred into that position. Next the old bubble is to be transferred out, and while this happens its replacement propagates one cycle further to the right. Two cycles of reverse rotation back up the new bubble and the minor-loop vacancy so that transfer-in can take place on the first cycle of forward propagation thereafter. This is not a true swap, because the old bubbles go into major-loop positions between the extrapolated positions of the new bubbles. This is of no importance, however, because after the secure write has been done the old block will be destroyed anyway. One can also use bidirectional propagation to speed up access to stored blocks. For a random block access we need $m/2$ steps to line up the block and $m/2$ more to detect it, in the BS/R organization. In the SML organization with bidirectional propagation we need $m/4$ steps to line up a randomly accessed block, on the average, and m steps more to transfer-out, detect and/or rewrite, and transfer-in. Thus on the average the time for a block of user data is only increased by 25% relative to the BS/R organization. The penalty might even vanish, or turn into a bonus, if locality of reference applies. The latter refers to the tendency of users to access together blocks that were stored together.

For multiple read or multiple write operations, however, there remains the slow-down by a factor of two (the time required to return bubbles to the transfer gates after detection). This penalty may even be larger, since chip design may dictate minor loops longer than four times the number of minor loops. In that case the added penalty can be avoided by returning the bubbles to the transfer gates after detection by reverse propagation. In this "shuttle" organization, the closed major loop would only circulate a block during a power recovery. The factor of two that remains, which is the time wasted returning the bubbles to the transfer gates without detection, might be recouped by combining different chips, or by using split circuits on one chip, so that user data flows in the detection channel during the whole time. The shuttle organization requires a detector that returns the bubble to the track in which it was detected.

3 ACTIVE COMPONENTS—TRANSFER

Several different ways to transfer bubbles from one ion-implanted propagate path to another have been considered. They can be grouped into four types that we call field effect, gradient effect, trapping, and reverse rotation transfer designs. The first three types use pulsed conductors, and descriptions of their operation overlap to some extent. We think of a field effect transfer as one that depends mainly on lowering the bias so that the bubble strips toward the intended position. A gradient effect transfer begins by pushing the bubble away from its initial position, which may otherwise compete for the bubble. In a trapping transfer gate the pulsed conductor merely holds the bubble within a specified region and the charged walls do the rest. Of course, in real designs the distinction between these three types is often not entirely clear and the grouping refers only to the major effect used. Reverse rotation transfer exploits the threefold symmetry in a gap oriented such that bubbles freely pass through it for one sense of rotation but propagate past as though it were a cusp when the field direction reverses. By forward and backward propagation through a succession of such gaps, bubbles can be manoeuvered in and out of the minor loops.

3.1 Field effect transfer

Starting with a bubble on a peak on one path, one can lower the bias and get it to strip out to a cusp on an adjacent track. The cusp generally has a stronger hold on the strip so that when the bias is increased again, the strip shrinks back to become a bubble there. To exploit this observation, we might place the minor loop ends opposite to cusps on the strong side of the major loop. A hair-pin conductor linking the minor loop end with the cusp could be pulsed to accomplish field effect transfer-out. We might even be able to get good transfer without the conductor, by pulsing the bias field (Lin and Sanders, 1981).

Satoh *et al.* (1982) described a bidirectional transfer design that operates by field effect. It could be used in the simple minor loop or shuttle organizations. Figure 14 shows propagation and conductor features and their orientation. The (111)-oriented substrate results in three preferred positions for a bubble on an unimplanted disk, as shown. As in the work of *Komenou et al.* (1981), the major loop was gapped to suppress nucleation by the transfer pulse. Bubbles propagate past the gapped region, in this orientation, much as they do past a cusp on an ungapped track. This fact has been used to design passive merge gates as described below. Because the transfer current raises the bias field at nearby minor loop positions, the end

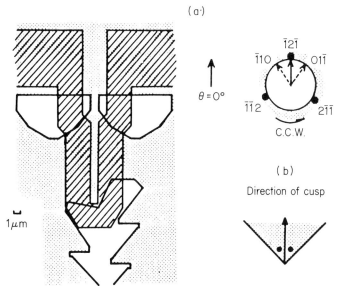

Fig. 14 (a) Cusp-to-cusp transfer design. (b) Definition of cusp direction. (After Satoh *et al.*, 1982).

of the minor loop was enlarged. To make room for it, the minor loop was folded. Expanded major-loop features were used to restore the usual two periods between transfer gates.

Altogether transfer-in and transfer-out each had 40-Oe bias ranges, or better, their margins were shifted relative to one another. Transfer-out limited overall operation at high bias, while the low bias margin for transfer-in was the same as the limit for minor-loop propagation. The overlapping bias range was about 30 Oe in a 100-kHz, 80-Oe triangular drive field. There were bubbles in neighboring minor-loop positions in these tests.

3.2 Gradient effect transfer

Consider again a minor-loop end opposing a major-loop strong side cusp. A conductor linking the two could be pulsed to cause a bias gradient from the one to the other, as envisioned by Bullock (1977). One problem with this is that the transfer-out pulse will lead to a bias elevation on bubbles in neighboring minor-loop positions. Furthermore, because the major-loop cusp is strong, it may take a large applied gradient to cause a bubble to transfer-in to a minor loop. However, gradient effect transfer-out can be made to work successfully. Nelson *et al.* (1980) were able to place the minor-loop end of their N-gate opposite a major-loop peak and still transfer

Fig. 15 N-Gate transfer design. (After Nelson et al., 1980.)

bubbles out to an adjacent major-loop cusp (see Fig. 15). This "peak-to-peak" arrangement does not degrade the low bias margin in minor-loop propagation.

The N-gate must normally be used with a transfer-in gate that inverts the data order. However, one variation of it (Nelson et al., 1981) extended the conductor as in Fig. 16. With this arrangement it was possible to pulse once to transfer bubbles out and again to move the bubbles over to the cusp on

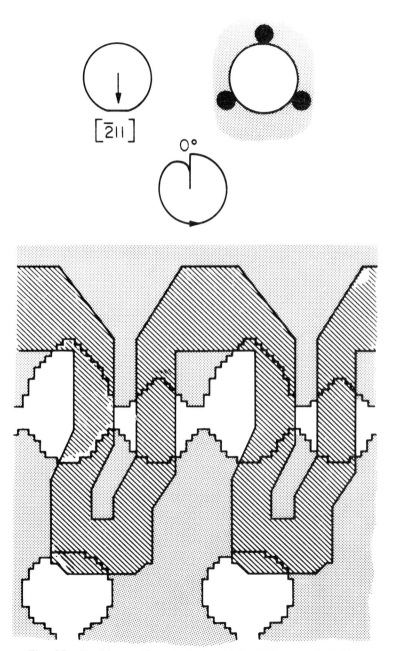

Fig. 16 Double transfer-gate design. (After Nelson *et al.*, 1981.)

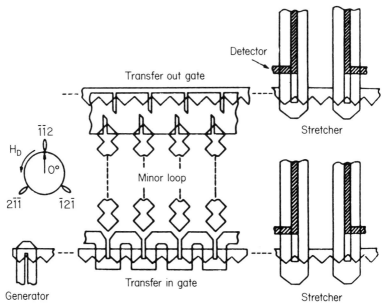

Fig. 17 Complementary transfer-gate pair. (After Komenou *et al.*, 1981.)

the roof-topped side. Since bubbles propagated from right to left there, the bit order was inverted. A noninverting transfer-in was used at the other end of the minor loops. Thus the N-gate can be used in a G-loop organization in two ways.

Komenou *et al.* (1981) included the N-gate in their 1-μm bubble device (see Fig. 17). Although they realized a 30-Oe bias range with it, better phase margins were obtained by using the transfer-in gate for transfer-out also. The tests were done at 300 kHz and required bubbles to move $\sim 10^3$ cm/sec. The transfer-in gate stripped the bubble from the gapped write line to the bottom of the minor loop, or vice versa. Note that the write line was gapped to suppress nucleation. The current required for nucleation increased from about 50 mA without gaps to about 175 mA with them.

The N-gate probably has the lowest electrical resistance of all. Its design was intended to apply pure gradient drive to the bubble coming around the end of the minor loop. Once free of the attractive edge of the end feature, the bubble should move towards the low bias region at the end of the slot extending downwards from the major-loop side of the conductor. At this time a charged wall extends from the major-loop cusp, and carries the bubble onto the track.

Komenou *et al.* obtained an overall bias range of 28 Oe at 50 ± 5 Oe drive, but the N-gate transfer-out had only 12 degrees of phase margins, and the effect of the transfer pulses on neighboring bubbles was not checked. This is of some concern for the N-gate because its current increases the bias field on neighboring minor loop positions. However, the current was only 50–60 mA and the conductor does not extend deeply into the minor loop, so the effect may not be significant.

Urai and Yoshimi (1982) used similar separate transfer-in and transfer-out gates in their 256-kbit device. They obtained 17 Oe overall bias margins with four-page bubble loading, but the frequency was a more normal 100 kHz, compared to Komenou's 300 kHz, and the cell size was $4 \times 4 \mu m^2$ instead of Komenou's $4 \times 5 \mu m^2$. The N-gate margins exceeded those for minor-loop propagation (10-page tests) at both ends of the bias range, so the transfer-out current cannot have had much effect on neighboring bubbles. The drive field was 60 Oe. This chip also included a single replicator, about which more will be said.

3.3 Trapping transfer

Figure 18 shows a bidirectional transfer gate design that defines a small "trap" between two opposing implant boundaries and a hair-pin conductor crossing them orthogonally. This design was based initially on the observation that 8-μm period minor-loop ends could be brought as close together as 3 μm without any loss in bias field propagation range. Transfer was achieved by pulsing the conductor to lower the bias in the trap, thus holding the bubble between the propagate paths for about 180°. During this time the attractive and repulsive charged walls exchanged positions, so the bubble jumps from one track to the other. The first attempt along these lines was by Jouve and Puchaska (1979), who used an ordinary propagation path opposed to the minor loop end. Excessive current was required at the normal spacing of one period peak-to-peak, and only transfer-in could be made to work. Unfortunately the spacing could not be reduced, as bubbles strip from the end of the minor loop into the major loop cusps at low bias. Since the minor-loop end could not be taken closer to the simple major loop, Nelson et al. (1981) added extensions to the major-loop peak to bring it closer to the minor-loop end.

Starting from two periods of 8 μm each between gates, Nelson et al. increased the width of the extended peak by 1 μm at the expense of the alternating periods. This makes the extended peak look more like a minor loop near the real minor loop end. Also, the bulge-to-bulge spacing was increased from 3 to 4 μm. The design has several advantages and a few remaining problems. Note that the conductor hair-pin opens toward the minor loop, and that the end feature of the minor loop is extended. Both

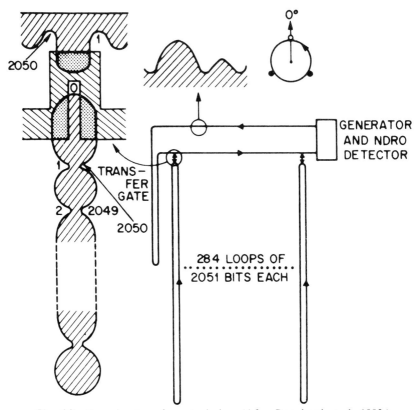

Fig. 18 Trapping transfer-gate design. (After Bonyhard *et al.*, 1982.)

improve transfer current margins by reducing the effects of the pulse on bubbles that may be nearby in the minor loops. The transfer-gate conductor has only a few squares per gate, keeping the resistance low in designs that may incorporate many hundreds of gates per chip. Also, the trapping current is not large. Figure 19 shows the current margins and illustrates that the low resistance per gate provides an attractive device for the design of system electronics. The phase margins, shown in Fig. 20, are also generally good. The only problem is the narrow trailing-edge phase margin in transfer out. The highest bias range is sometimes obtained only in a narrow range near the latest trailing-edge phase. On the other hand the bias range for transfer-out is sometimes limited only by major-loop propagation bias over a wide range of trailing-edge phase. The factors contributing to this variability seem to be process-related and are still under study.

In the most recent version (Bonyhard *et al.*, 1982), the major-loop extension was widened again, resulting in a minor loop to minor-loop spacing

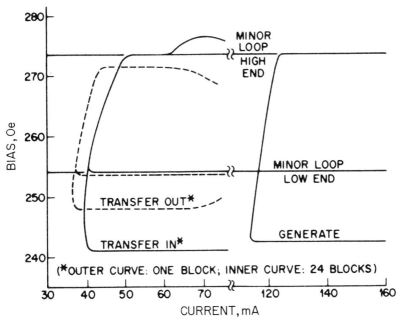

Fig. 19 Trapping transfer current margins. (After Bonyhard *et al.*, 1982.)

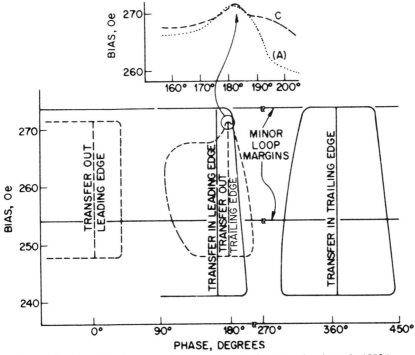

Fig. 20 Trapping transfer phase margins. (After Bonyhard *et al.*, 1982.)

of 18 μm. The minor loop period was reduced to 7 μm to compensate for the increased major-loop period. To improve transfer-out phase margins, the conductor slot was angled toward the side of the extension that the bubble should occupy after 180°. This design also worked well when scaled down by a factor of two to (nominally) 4-μm period (Nelson et al., 1983).

3.4 Reverse rotation transfer

Nelson et al. (1980) discovered that narrow gaps between unimplanted disks can have asymmetrical propagation properties. Figure 21 shows that bubbles propagating on the strong side of a gapped track find that the gap acts like a cusp, but bubbles approaching from the other side pass freely through it. Thus gaps can be used to merge bubbles propagating on separate paths into one path.

It seems to be necessary to restrict the gap width to $\lambda/4$ to avoid losing high bias operation (Nelson et al., 1980). This is unfortunate because the resolution in ion-implanted circuits can otherwise be coarsened to $3\lambda/8$, but the gaps are useful. We have already seen them used to suppress nucleation (Komenou et al., 1981; Satoh et al., 1982; Urai and Yoshimi, 1982). Moreover gaps can be used to invert the bit order after transfer-out. Figure 22 illustrates

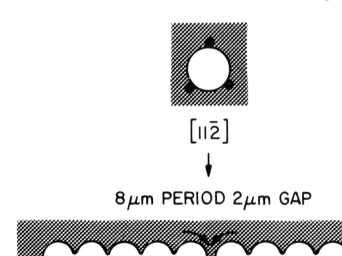

Fig. 21 Rectifying action of properly oriented gap in ion-implanted propagation patterns. (After Wolfe et al., 1981.)

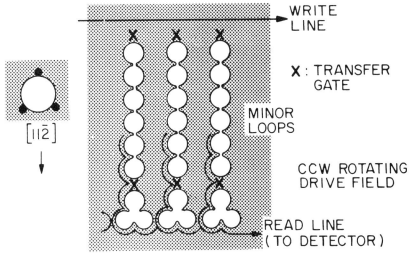

Fig. 22 Use of a gapped read line to invert data order after transfer-out. (After Wolfe *et al.*, 1981.)

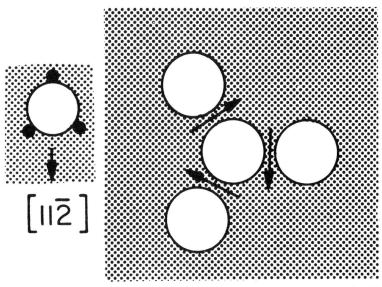

Fig. 23 Triplet of merge gates as determined by threefold symmetry. (After Wolfe *et al.*, 1981.)

this with a gapped-read line made up of clover-leaf-shaped features. Although not shown in detail, transfer-in from the write line to the minor loops and transfer-out from the minor loops to the clover leaves could both be done with trapping transfer gates.

Once bubbles have merged, reversal of the propagation direction keeps the bubbles exclusively on the strong side path. Because of this "rectifying" effect, transfer gates can be made that operate by a combination of forward and backward propagation steps through two such merges. Since reverse rotation transfer (RRT) gates use no conductors, there are no crossings that could interfere with propagation (see Horng and Schwenker, 1981 about this subject) and no increased system voltage requirement as might be the case with a series connection of hundreds of transfer gates. Also, the

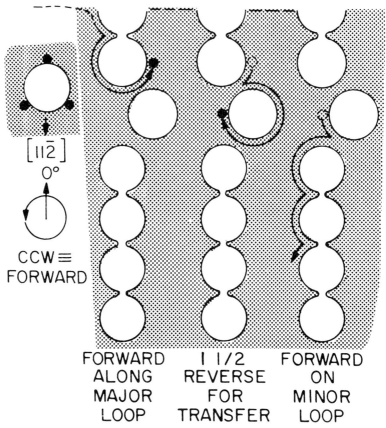

Fig. 24 Bidirectional transfer gate using merge gates with reversed rotation and no pulsed conductors. CCW, counter-clockwise. (After Wolfe et al., 1981.)

problems of determining pulse amplitudes, positions, and widths are eliminated. Although a practical system would be complicated by extra logic in its drive field control section, it would also be simplified by the removal of one or more current sources, switches, and timing generators. The merge behavior is seen whenever the gap is aligned with a $\bar{2}11$ crystal axis and the gap width is comparable to the bubble diameter. There are three such orientations, as illustrated by Fig. 23, and they can be combined to advantage in the reverse rotation transfer gate to be described below. Note that gaps oriented at $90°$ to a $\bar{2}11$ axis do not impede the motion of bubbles in either direction. Indeed, the success of the trapping transfer strategy depends on the smooth propagation of bubbles through such gaps.

Figure 24 shows a simple bidirectional reverse rotation gate (Wolfe et al., 1981; MacNeal and Gergis, 1981) and shows in three steps how $\frac{3}{2}$ reverse rotations, starting at $270°$ and ending at $90°$, can be used to transfer a bubble from an upper loop around an "idler" disk to a lower loop. Transfer in the opposite direction is similar with reverse rotation starting at $90°$ and ending at $270°$. It is easily verified that in the RRT process the neighbors of the transferred bubble suffer no net disturbance. In an obvious application of the RRT gate, the upper loop (for example) may be thought of as part of a major loop in a major–minor loop organization, while the idler and lower loop represent one each of many transfer gates and minor loops, with all transfer gates controlled simultaneously by drive field rotation reversal. A useful property of this RRT gate is that when a transfer-out reverse rotation sequence is followed immediately by a transfer-in sequence, the bit that is removed from the minor loop is replaced by a new one from the major loop. This is not a true swap, however, since the bit from the minor loop is placed in the major loop one rotating field cycle ahead of the replacement bit.

4 ACTIVE COMPONENTS—GENERATION, REPLICATION, AND DETECTION

Besides transfer gates, a minimal chip layout needs a generator and detector. Since the practicality of replication in ion-implant circuits is not yet established, nondestructive detection must be considered also.

4.1 Generation

Relatively little has been written about generator design in ion-implanted circuits. Rather, early designs like that shown in Fig. 25 have been used

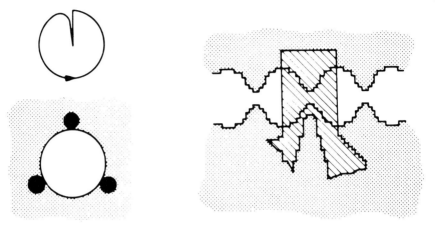

Fig. 25 Bubble generator structure comprising a hair-pin conductor located over a strong side cusp. (After Nelson *et al.*, 1980.)

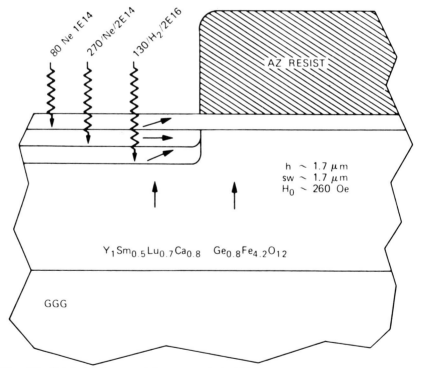

Fig. 26 Triple implant used for magnetic bubble propagation. The shallow dose was unpatterned and helped increase the nucleation threshold. (After Nelson *et al.*, 1980.)

without much modification. The hair-pin conductor was placed over a strong side cusp because early devices sometimes were observed to nucleate bubbles there spontaneously. The conductor slot was stopped short of the implant boundary on the assumption that bubbles might otherwise be generated inside the unimplanted area. Nelson *et al.* (1980) hit on the expedient of using a uniform implant to increase the threshold current for nucleation (see Fig. 26). Although this stopped spontaneous nucleation, the minimum generator current was still low for low anisotropy samples, and passivated samples needed even less current than unpassivated ones, indicating conductor stress involvement.

The uniform implant was intended to suppress hard bubbles in the detector, where bubbles were stripped out in otherwise unimplanted material. The uniform implant became necessary to suppress nucleation by control function other than the generator. For example Komenou *et al.* (1981) found their transfer-in gate nucleated at 20 mA without the uniform implant and at 50 mA with it. This was still too low, so they introduced gaps in the propagate path. In their improved design, the nucleation threshold was about 175 mA. On the other hand, Asada and Urai (1982) found better generator current margins with double He implantation and no uniform implant than with double H_2 implantation and a uniform dose of 10^{14} Ne^+/cm^2 at 50 keV. The current threshold was higher with the uniform implant included, however. The maximum current was limited by extra bubble errors. In both of their samples, Asada and Urai found acceptable operation over a temperature range of 20–90°C.

In permalloy devices, it is thought that nucleation takes place at the surface of the bubble film in the intense fields of the closely spaced conductor (Nelson *et al.*, 1973), but in ion-implanted circuits the effective spacing is larger by the thickness of the implanted layer. This is typically about 0.5 μm (over which the fields decrease significantly), so why do generators work so easily in ion-implanted circuits? Almasi *et al.* (1980) concluded that the charged wall contributed about $2\pi M$ in opposition to the bias field in their experiments.

We assume that nucleation takes place when the condition

$$H_x^{2/3} + H_z^{2/3} \geq (H_K - 4\pi M)^{2/3} \tag{7}$$

is satisfied somewhere in the bubble storage layer. This condition defines a critical curve, or asteroid, in the H_x, H_z plane. Now there are two stable directions of the magnetization for points inside the curve, but there is only one solution otherwise. There will therefore be combinations of H_x and H_z that can overcome the anisotropy and invert the magnetization. Figure 27 shows the sources acting to cause nucleation near the implant boundary. Besides the charged wall, there is the thickness step at the implant boundary,

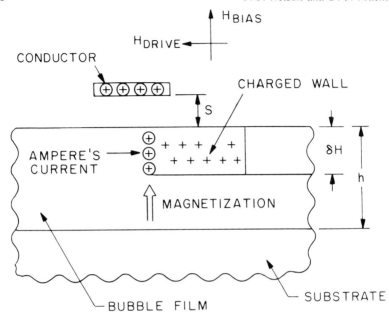

Fig. 27 Schematic diagram of magnetic field sources acting to generate bubbles in strong side cusp of I2P2 device.

which can be thought of as carrying an Amperian current $cM\,\delta h$. The bias field opposes nucleation, but the drive field adds to the in-plane components of the charged wall and Amperian current. Finally, there are the intense fields caused by the pulsed conductor. The effect of the pulse will be greatest where the in-plane and vertical components are roughly equal, because the combination is $2^{3/2}$ (about 2.8) times larger than one component acting alone. Also, the two components vary rapidly with position near the conductor, while each tends to be much smaller where the other goes through its maximum. Almasi et al. attempted to account for nucleation using the bias field, the vertical component of the charged wall's field, and both components of the field caused by the pulsed conductor, but they neglected the in-plane field of the charged wall, both components of the field due to the magnetization discontinuity, and the drive field, all of which also aid nucleation. In any case, the assumption of a two-dimensional charge distribution is unrealistic because the field at points near it diverges. Almasi et al. took the field points to be 0.1 μm below the charged wall, but we know of no justification for this procedure. It seems that an accurate accounting of generation will have to wait for more detailed modeling of the magnetization distribution.

4.2 Replication

Besides true NDRO, in which the bubble is returned after detection, bubble circuit designers have employed "replication" followed by DRO of one of the copies of the original bubble stream. In permalloy devices the following strategy works: (1) a bubble passively expands along a large feature's edge, (2) a hair-pin conductor lying orthogonally to the edge is pulsed to cut the stretched bubble in the center, and (3) poles near the outside edges of the conductor catch the two bubbles. On ion-implanted propagation patterns, however, bubbles do not expand along the feature edges. If any expansion does occur, it is along the charged walls, usually perpendicular to the track edge. By accident, Nelson et al. (1981) found that replication can be accomplished by the opposite strategy—stretching the bubble actively with a conductor followed by passive cutting by negative charged walls.

Figure 16 showed an 8-μm period "double transfer" gate. The rounded features at the bottom are minor-loop ends. Bubbles propagating around the ends could be transferred to the major-loop cusps above and slightly to the right by a pulse starting before about 330° and lasting until about 350° or later. This is N transfer-out, really. But the conductor had legs that were extended to the back or "roof-topped" side of the major loop. Following transfer-out, a second "invert" pulse was applied to move the bubbles over to the upper track. In this invert transfer, a bubble was stretched between two cusps, and at the end of the pulse the first cusp was no longer able to hold its end of the strip. By the way, Sanders et al. (1982) also described an invert transfer gate. Their gate transferred bubbles into the minor loops but acted by the gradient effect.

The invert pulse used by Nelson et al. required a current of 80 mA and a trailing-edge phase later than about 148°. Figure 28 shows what was discovered when the leading edge phase was varied with the current increased to 100 mA. The symbols R and T stand for replicate and transfer, respectively. For early leading edges, negative charged walls formed in the strong side cusp apparently divided the stripped out bubbles. When the pulse started late, the bubble probably avoided the charged wall as it stripped out to the cusp on the roof-topped side. A similar design was subsequently used by Bonyhard et al. (1980) as a point replicator feeding a DRO detector. Such replicators did not prove to be reliable, however, and NDRO was thought to be a better approach.

Urai et al. (1982) added an extra conductor level to their device. That is, the stretching and cutting were both active. While this seems straightforward, its operation was not as simple as it might sound. The cut pulse should not overlap the stretch pulse in time, because excessive cut current would be required to reverse the stretch field. Therefore the propagation design must

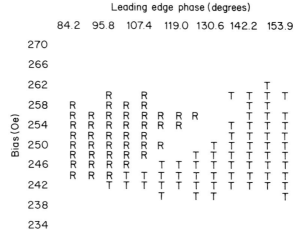

Fig. 28 Bias margins versus leading edge phase for transfer (T) and replicate (R) modes of operation of double transfer gate. (After Nelson et al., 1981.)

be capable of holding the stripped end of the bubble after the stretch pulse terminates, so one of the target positions was made a strong side cusp. But then the replicator nucleated bubbles for some phases. In spite of these difficulties, a 256-kbit chip was operated with two-level replication over a 12-Oe bias range (Urai and Yoshima, 1982).

Now strong side cusps contain attractive charged walls longer than

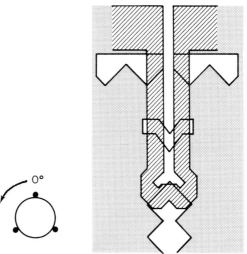

Fig. 29 Block replicate gate formed by interposing V-shaped unimplanted feature between minor-loop end and major-loop gap. (After Komenou et al., 1982.)

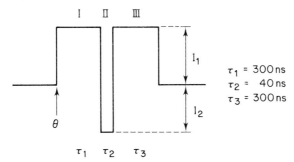

Fig. 30 Compound pulse used in block replication. (After Komenou et al., 1982.)

repulsive walls. Therefore a better replicator design might try to cut in a weak side cusp where repulsive charged walls are favored. Figure 29 shows a design tried by Komenou et al. (1982). This design had 20 Oe of bias range at an error rate of 10^{-5} when tested with an 80-Oe (peak) triangular drive field at 200 kHz.

Note that the major line was gapped as before (Komenou et al., 1981; Satoh et al., 1982). A V-shaped unimplanted feature was interposed between the minor loop and major loop. Its weak side cusp was designed to provide a repulsive charged wall to cut the strip. Without the V-shaped feature, bubbles could be transferred-out but they could not be replicated.

Figure 30 shows the somewhat unusual replicate pulse that was used. It can be thought of as a positive stretch pulse I to which a short cut pulse $(I_1 + I_2)$ was added. Replication was more sensitive to the leading edge phase θ, the stretch current I_1, and the cut width τ_2 than to the other parameters. For a 20-Oe bias range there was a 14° phase margin in θ, a 10-mA current margin for I_1, and a point margin at 40 nsec for τ_2. Shorter cut pulses might have worked even better, but no data was given.

4.3 Detection

Detection, like replication, is made difficult in ion-implanted circuit design by the lack of a passive stretching element. The obvious alternative, actively stripping the bubble with a pulsed conductor, has been used successfully. Naturally there is a limit to the detector length that is set by the saturation velocity (de Leeuw, 1978) and by the time available for stretching. This works out to be on the order of 100 μm, which is much less than the detector length in permalloy bubble devices. To increase the signal amplitude, thin permalloy (300–500 Å) has been used for sensing. The resulting structures look much like those used for the same reasons by Copeland et al. (1973)

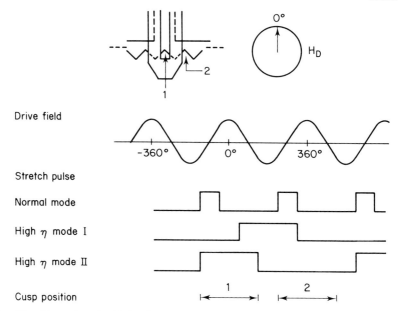

Fig. 31 Normal and "high efficiency" η modes of operation of bubble detector using active stretching. (After Komenou et al., 1981.)

in a conductor access bubble device. There is also the problem of what to do with the bubble after detection. After the stretch pulse ends, the strip begins to contract. After a delay, it is possible to clear the detector with a reverse-polarity pulse applied to the stretch conductor. It is also possible to position one end of the strip so that the bubble returns to a propagation path.

Komenou et al. (1981) investigated the problems of high-frequency detection in a 1-μm bubble device. Figure 31 shows stretch pulse phasing for the normal mode and for their high efficiency mode. Mode I gave better signal versus bias than mode II. In mode I the bubble was stretched during the time it would normally have propagated to the next cusp past the detector. The point of stretching for more than $\frac{1}{2}$ of a cycle was, of course, to give the bubble more time to expand. Improved current margins were obtained by stepping the stretch current down to 50 mA after 0.2 μsec of the 2-μsec-wide stretch pulse. This implies that higher drive was needed to initiate the expansion across the implant boundary. The permalloy sensor was 2 μm wide, 100 μm long, and 300 Å thick. It was centered in the conductor opening and a signal of 8 mV over a 40-Oe bias range was realized at 300 kHz.

Of course, the limiting factor in these detectors is heat. Urai and Asada

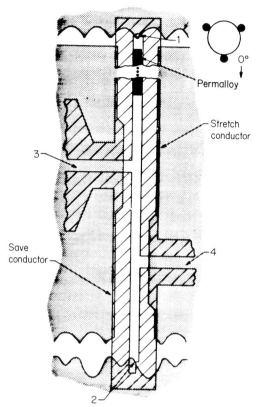

Fig. 32 NDRO detector design used on 8-μm period ion-implanted bubble circuits. (After Ekholm *et al.*, 1982.)

(1982) used the change in permalloy sensor resistance to estimate the temperature rise. A temperature rise of 8°C was then used as one boundary of the operating area on a plot of stretch current versus width. The other two sides of the safe area were given by the minimum current, ~60 mA, and the minimum pulse width, ~3 μsec, needed to realize 6 mV signals. Fortunately, these limits enclosed a positive area. Urai and Asada did not give the detector bias current, although it may have contributed significantly to the temperature rise. They used a temperature coefficient of 0.25%/°C, which is significantly different from the 0.4%/°C used by Marsh (1938). Therefore the 8° temperature rise may have to be corrected to 13°.

In the work mentioned above, bubbles were collapsed after detection. This is fine in a chip design using replication, but if replication is unavailable then it is better to reuse the bubble than to use the detected output to regenerate the information. Figure 32 shows a practical NDRO structure

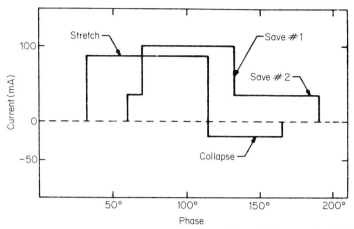

Fig. 33 Pulses used in NDRO detector. (After Ekholm *et al.*, 1982.)

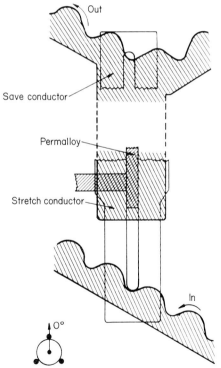

Fig. 34 NDRO detector design allowing about $\frac{1}{2}$ cycle after detection for destretching. (After Bonyhard *et al.*, 1982.)

described by Ekholm *et al.* (1982) for 8-μm period devices. The bubble was stretched from a strong side cusp as in the other work, but the stretch conductor also carried the strip past the permalloy sensor to be picked up by a second conductor circuit. This "save" conductor was pulsed to hold its end of the strip while a low collapse pulse was applied to the stretch conductor (see Fig. 33). Finally the strip had to contract to a bubble on the "roof-topped" outgoing track on its own. Error rate limits as low as 5×10^{-9} were determined at 50 kHz. For 100-kHz operation, this design was modified to that shown in Fig. 34. In this case less than about 90° of stretch current were needed to strip the bubble along the detector, but after sensing at

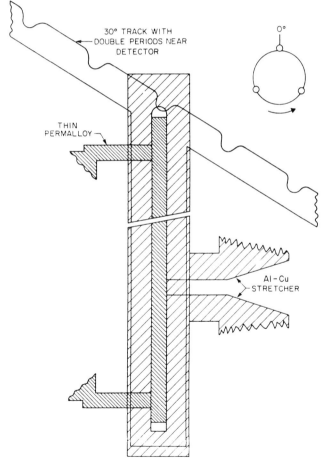

Fig. 35 NDRO detector design that returns bubble to cusp of track from which it is expanded. (After Nelson *et al.*, 1983.)

$-90°$ the save pulse was maintained for over half a cycle. The extra time provided for more reliable de-expansion. Finally, the strip became a bubble on the 30° track some time after 120°, when the target cusp was most attractive.

It was also proved to be possible to stretch the bubble from the cusp to which it is to be returned. Figure 35 shows the arrangement used by Nelson et al. (1983) for NDRO detection at 50 kHz in a 4-μm period device. The sensor was 375 Å thick, 1.5 μm wide, and 56 μm long. It produced 10-mV signals with 3 mA of bias current. Bubbles were nondestructively detected with a single stretch pulse that lasted about 50° and ended around 120°.

5 PROPAGATION

As might be expected in (111) films, the interaction of bubbles with unimplanted circles exhibits threefold symmetry, and the Bitter technique shows planar domains in the implant, sometimes separated by charged walls. These charged walls, which carry bubbles along the propagate pathways, are governed by the cubic and uniaxial magnetic anisotropies as well as by magnetic fields. Stress relaxation and the demagnetizing field at the pattern boundary are especially important but difficult to calculate. However, the effects of the three-fold symmetry on straight run propagate tracks are known empirically. Fortunately, a good orientation exists for minor loops, which carry propagation paths that must run in two opposing directions. The strongest propagation direction remains available for one side of the major loop. A good choice for the start–stop direction, affording protection to all bubbles in minor loops either by cusps or by preferred positions at the loop ends, is also available.

5.1 Track orientation

The bias margin range for contiguous-disk propagation in different directions exhibits threefold symmetry. Figure 36 shows margins versus drive field for propagate tracks parallel and perpendicular to an easy–hard axis (Wolfe and Nelson, 1978). Of course, back-and-forth propagation is the same for the two sides of the vertical loop, since the easy directions and the track details are both mirror symmetric through the axis of the loop. However, the two sides of the horizontal loop are inequivalent with respect to the threefold symmetry. One side is called "super", and the other "bad", compared to the "Good" vertical propagate paths. The Ferrofluid pictures

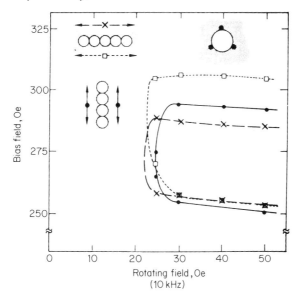

Fig. 36 Asymmetric propagation of 1.7-μm bubbles on 8-μm I2P2s in cage material with Ne–H$_2$ implantation. (After North et al., 1978.)

in Fig. 37b, taken with no in-plane field acting, show unequal attractive and repulsive charged walls on the super and bad tracks (in this figure the right and left sides of the vertical track). The super track has net attractive (black) charged wall, while the bad track is repulsive (white). In both cases the stronger wall is formed in the cusp by magnetization directed along two of the easy directions. At the same time, the two magnetization directions lie nearly parallel to the track edge, which is therefore nearly neutral. That is, the magnetization appears to choose threefold easy directions and to avoid large demagnetizing fields. Weaker charged walls form at the peaks, while the charged walls in the cusps appear not to extend out beyond the peaks. Further from the tracks, uncharged walls run parallel to a line joining the peaks. There both directions of magnetization may join to a surround formed by either solution. On this basis, the wall emerging from the cusp is longer than one attached to a peak by a distance equal to the track modulation from peak to cusp. Similarly, in Fig. 37a the two good tracks (horizontal here) have equal black and white charged walls. And an uncharged domain pattern forms that joins the magnetization in the cusps to an outer magnetization that is also one of the preferred directions.

There are three bad tracks and three super tracks, separated by 120°, and six good tracks separated by 60°. Our characterization applies to combined propagation in clockwise and counter-clockwise rotating fields. If

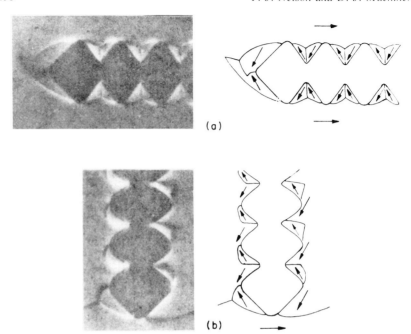

Fig. 37 Ferrofluid patterns showing different domain wall structures associated with (a) good, and (b) bad, and super tracks. (After Dove and Schwarzl, 1979.)

only one sense is used, then the six good directions split into two groups of three separated by 120°. The shuttle organization requires loops that propagate bubbles in opposite directions, and efficient use of material demands the minor loops enclose minimum area. Obviously the two oppositely directed good tracks, propagating parallel to an easy–hard axis, offer some advantages for use in a minor loop. First, of course, is that the bad direction can be avoided in the storage loops, where most of the bubbles circulate most of the time. Moreover the minor loops so formed will propagate bubbles equally well in clockwise and counter-clockwise rotating drive fields. This is an important consideration for the shuttle organization or for reverse rotation transfer. Also, the end turns are easy to design, since portions of disks will do. On the other hand, Lin *et al.* (1979a) suggested using tracks at 15° to the easy–hard axis. This has the attractive consequence that the major loop, lying at 90° to the minor loops, shares their orientation relation to the crystal axes. However, the two sides of such loops are not equivalent in any sense. This idea does not seem to have been pursued, so we consider instead the design of minor loops running parallel to an easy–hard direction.

5.2 Track design

The original "contiguous disk" was conceived of as a gapless, low-resolution design, but the resolution required for its formation was not spelled out. Later track designs were introduced with "cusps" that had finite positive curvature. Their resolution can be defined as twice the minimum radius of curvature along the track edge. The resolution required for a permalloy propagate design would be equal to the gap between elements or the smallest feature linewidth.

Lin et al. (1977) advocated sine waves on the premise that lithography would distort such tracks less than any other design. However, the radius of curvature in the trough of a sine wave is rather small. Nelson et al. (1979) have used cusps that are more rounded out, so that the resolution required is $3\lambda/8$, with good results. (Their peak-to-valley height is also $3\lambda/8$.)

Besides the contiguous disk minor loop design, snake-shaped loops have been used. As shown in Fig. 38, the snake track shape is not different from contiguous disk, but the relation of the tracks to each other was modified. Snake loops have wider implanted regions between loops. According to the model of Best (1981), this should give lower drive field thresholds. Since the low drive field limit is experienced as a loss of low-bias margin, snakes are used to improve low-bias operation, too.

In another development, the triangular track elements shown in Fig. 39 were considered (Shir, 1981). A bubble arrives at a peak when the drive field points in an easy direction of the threefold anisotropy. This minimizes the length of the charged wall there and the chance that a bridge will form to the neighboring track. That the low bias margin is improved is supported by the data of Ju et al. (1981). On the other hand, the effect on the minimum drive field is not clear, since the edges of the propagate path are deliberately turned away from the easy magnetization directions. Indeed, Wolfe and Nelson (1978) found improved bias margins and lower minimum drive fields for minor loops made of threefold elements turned oppositely to Shir's. While improved understanding of the interplay between threefold anisotropy, stress relaxation, and demagnetizing effects may yet lead to better track design, at present the original contiguous-disk design seems to be as good as any.

The next complication in minor loop design results from folding. Folded minor loops leave room for large structures in the gate regions and are becoming standard in permalloy designs. Further, folded ion-implanted minor loops would be needed in hybrid circuits, which use permalloy gates. Also, fewer gates mean lower resistance, so that less voltage needs to be supplied by the support circuitry. However, folded loops are slower, being larger, and also require additional turn designs. Satoh et al. (1982) developed

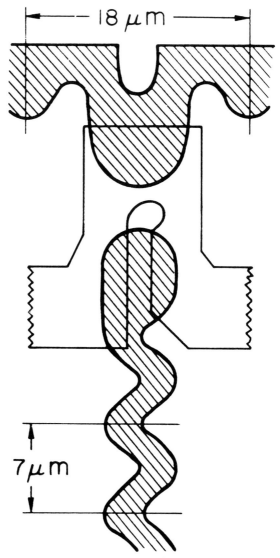

Fig. 38 Snake minor loop design for $7 \times 9\,\mu m^2$ cell dimensions. The far turn was a copy of the end shown, but rotated by 180°. (After Bonyhard et al., 1984.)

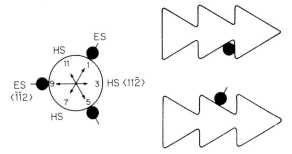

Fig. 39 Triangular-shaped propagation patterns and their orientation with respect to the cubic crystalline anisotropy. (After Shir, 1981.)

the 180° inside turn shown in Fig. 40. In this design the turn itself is in the weak track orientation. Therefore the cusp was tipped (with respect to the bisector) by 30°, which coincides with one of the medium track directions. Good propagation at 4-μm period was obtained.

The use of gapped strong sides tracks in the major loop has been discussed. Bubbles honor the gaps as regular cusps, but the track resolution has to be tightened to $\lambda/4$. The gaps are useful either for suppressing nucleation or for rectifying data order after transfer-out. The strong side has also been modified to aid transfer. Good propagation was achieved with extended periods that carry bubbles into the region of trapping transfer gates. The

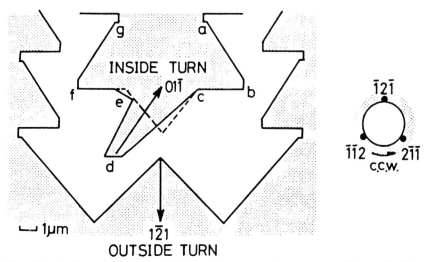

Fig. 40 Inside turn design using triangular propagation patterns. The weak side is avoided by tipping the cusp in the middle of the turn by 30°. (After Satoh et al., 1982.)

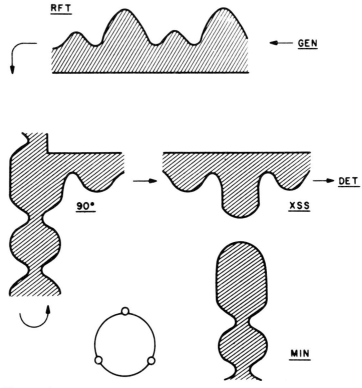

Fig. 41 Roof top (RFT), track leading from generator (GEN), 90° inside turn, extended strong side (XSS), minor-loop (MIN) design and track leading to detector (DET) as used in 8-μm period ion-implanted devices. (After Nelson et al., 1982.)

extenders take up half of the distance that existed between the minor loop ends and regular tracks, which was λ.

It is advantageous to try to avoid weak side propagation, but in a chip design without replication the major loops have to be closed. One could use tracks at 30° and 60° to the weak side, but a large area would be enclosed. Wolfe and Nelson (1979) obtained improved propagation on the "roof-topped" track shown in Fig. 41. These were symmetry-inspired features that propagated bubbles at $\pm 30°$ to the weak side in alternate periods.

The simple major loop design used by Bonyhard et al. (1982) also required a 90° inside turn to connect a count-adjusting minor-loop-like appendix to the strong side. As shown in Fig. 41, this looks something like an extended period joined to a medium side. Both the roof top and the 90° turns were used in 8-μm devices. The roof-top track has been replaced by the 30-30 track of Bonyhard et al. (1984). Here, $\pm 30°$ tracks were used explicitly and were connected at peaks directly and at valleys by a simple extension (see Fig. 34). The 30-30 track has given good results at 4-μm period, but the 90° inside turn did not work as intended (Nelson *et al.*, 1983). In fact, at 4-μm period, bubbles skipped the last cusp before the strong side. Although they propagated reliably in this short count mode, low end bias was lost and reverse propagation through the turn was not possible.

6 CHARGED WALLS

The formation of charged walls at implant edges is known as the result of recent advances to be due largely to the phenomenon of stress relief at the boundaries, and this effect is therefore ultimately crucial to bubble propagation. We will here discuss in some detail the simplest possible case, the propagation of a bubble around a single unimplanted disk.

Despite the analogy with permalloy propagation patterns that led to the discovery of bubble propagation on structures formed from patterned implants, the propagation mechanisms are very different in the two cases. Wolfe and North (1974) first observed planar domain walls in the implanted layer radiating out in the form of spiral arms from unimplanted disks using the Ferrofluid Bitter pattern technique. Almasi *et al.* (1974) pointed out that these walls carry magnetic charge and are the agents that carry bubbles along in patterned implant propagation.

Almasi *et al.* considered the in-plane magnetization in the implanted layer to flow around an unimplanted disk "much like a slow stream of water flowing around a boulder, forming a diverging wall upstream and a converging wall downstream" (Fig. 5). Unlike the case with normal Bloch and Néel walls, the magnetization divergence in these walls is nonzero, with opposite polarity on the two sides of the disk. For a bias field pointing up out of the garnet surface, the top of a bubble will be negatively charged, and so it will be attracted to the positively charged wall where the magnetization converges on the "downstream" side of the unimplanted disk. The direction of the inplane magnetization at a distance from the disk can be controlled by an applied field in the film plane, and so the charged walls as well as any attached bubble can be moved around the disk by a rotating drive field.

When observed using Ferrofluid (Fig. 42), positively charged walls appear darker than the background and are hence sometimes referred to as "black walls", while bubbles and negatively charged (or "white") walls appear lighter. This is because the dark magnetic particles in the Ferrofluid are oriented by the dominating bias field to have their negative poles directed toward the garnet and so are attracted to the positive charged walls and

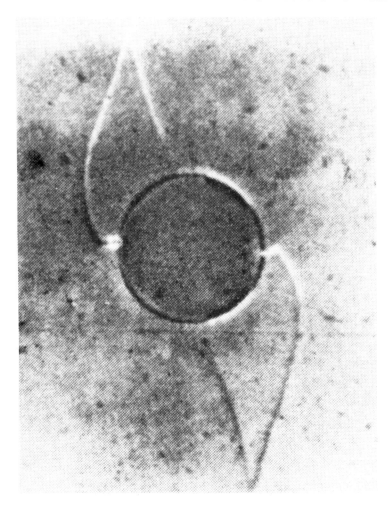

Fig. 42 Micrograph of the typical "propeller" domain pattern around a 20-μm-diameter unimplanted disk with a 20-Oe in-plane field to the left. A white bubble lies under the attractive black wall on the left of the disk. (From Lin et al., 1978.)

repelled by bubbles and negative walls. If the bias field is reversed (that is, directed downward into the bubble film), it is still the black charged walls that attract bubbles. In this case, naturally, bubbles are positively charged on top and black walls carry negative charge. We will assume in general that the bias field points up, although it makes no difference to the physics of processes we consider. Even uncharged or conventional walls, at least of the Néel type, may still have enough stray field to make them visible.

Ferrofluid photographs showing domains in implanted layers abound in the literature.

While several papers have been published on the internal structure of charged walls in implanted garnet films (Hubert, 1979; Shir and Lin, 1979; Kleman and Puchalska, 1980), a detailed discussion of this subject in its present state would not help us much here.

The picture given above of bubble propagation around the simplest of all unimplanted patterns, the disk, was considerably oversimplified. From the first observations of the behavior of bubbles around unimplanted disks (Wolfe et al., 1971) in (111) films, a threefold symmetry was evident. In the absence of an applied inplane field, a bubble on a disk comes to rest at one of three positions spaced at 120° intervals around the disk, and if a moderate in-plane field drives it around the disk, the bubble slows down at these favored positions and speeds up in between. Ferrofluid observations of domain structure in the implanted layer also revealed patterns indicating an underlying threefold symmetry.

An obvious effect that exerts an influence with threefold symmetry on the magnetization in the implanted layer is cubic magnetocrystalline anisotropy. The cubic term in the anisotropy energy is given by $E = K_1(\alpha_1\alpha_2 + \alpha_2\alpha_3 + \alpha_3\alpha_1)$ where the α terms are the direction cosines of the magnetization with respect to the crystal axes. For negative K_1, as is usual in bubble films, the (111)-equivalent directions are easy axes of magnetization. While for magnetization exactly in the (111) plane there is no threefold dependence of anisotropy energy on azimuthal angle, if the magnetization is tipped out of the plane by the usual bias field applied in the (111) direction, it will pass closer to cubic easy axes when its in-plane projection is along the three $[-2,1,1]$ directions than when pointing along the opposite $[2,-1,-1]$ directions. In this way we have three effective easy (and three hard) directions in the (111) plane when a perpendicular bias field is present.

Shir and Lin (1979) and Callen (1979) obtained an expression for the threefold anisotropy energy,

$$K_3 = \cos(3\phi) = -\frac{\sqrt{2}}{3}\frac{K_1 M H_b}{K_1 + 2K_u - 4\pi M^2}\cos(3\phi) \tag{8}$$

where ϕ is measured from a $[-2,1,1]$ in-plane easy axis. The effective threefold anisotropy field (equal to the minimum applied in-plane field that can saturate the in-plane magnetization in all directions) is given by $H_{K1} = 9K_3/M$. Then K_u, the net uniaxial anisotropy resulting from the combined effects of growth and implantation, is of course negative and assumed large enough that the bias field produces only a small out-of-plane angle ($\sim 15°$ in the calculation of Shir and Lin). Although K_1 is independent of the applied bias field H_b, H_{K_1} depends on and is proportional to H_b. The threefold anisotropy energy dominates the determination of in-plane domain and charged wall configurations far from strongly perturbing influences such as unimplanted propagation tracks.

Naturally we are most interested in understanding the behavior of in-plane domains and charged walls in the neighborhood of not only implant patterns but also magnetic bubbles. This makes the problem enormously more complicated. One way in which the implant edge makes itself felt on the magnetization field is due to the divergence of in-plane magnetization at the pattern edge. The demagnetization field tends to make the in-plane magnetization orient itself parallel to the edge as in the "flowing water" analogy of Almasi et al. (1974). Until recently, in fact, the combination of magnetocrystalline anisotropy and the demagnetization field were believed to govern the behavior of charged walls near unimplanted patterns. Now it is understood that in fact charged walls will not even form around an unimplanted disk (Hidaka and Matsutera, 1981), except for the anisotropy due to the partial relaxation of the implant induced stress at the pattern edge. Further, the directional dependence of the inverse magnetostriction effect that generates a magnetic anisotropy from the stress relief, rather than the magnetocrystalline anisotropy energy K_1, accounts for much of the threefold symmetric behavior of charged walls around a disk (Hubert, 1984) and by extension the direction-dependent properties of propagation tracks discussed earlier in this chapter.

Hidaka and Matsutera (1981) studied charged-wall formation in various magnetic garnet layers with net negative uniaxial anisotropy and hence lying-down magnetization. Charged walls were formed (or not formed) at the edges of either unimplanted patterns or actual holes ion-milled into the films. What they found to be the essential element for the formation of charged walls at pattern edges was not cubic anisotropy but compressive stress in the film. High cubic anisotropy without compressive stress did not lead to charged walls; compressive stress with nearly zero cubic anisotropy did give charged walls.

The magnetostriction constant λ_{111} in Hidaka and Matsutera's films was negative, as is usually the case in (111) bubble films where ion implantation is used to create a compressive stress that lays the magnetization down in the implanted layer. They pointed out that material in the unimplanted pattern in these cases is not under compression, and that this leads to stress relief at the boundary. Figure 43, which is similar to that of Hidaka and Matsutera, illustrates this process. At the pattern edge the implanted layer (cross-hatched in the figure) pushes in on the unimplanted part of the film, putting it under local compression perpendicular to the edge and reducing

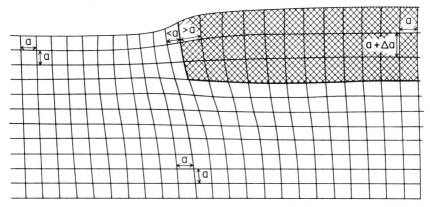

SLICE ⊥ TO IMPLANT EDGE

Fig. 43 A cross-section of an implanted film perpendicular to an implant edge. The implanted volume is shown cross-hatched. Far from the edge in the implanted region, the lattice is strained only in the vertical direction. A thick substrate constrains the implanted material from expanding in the horizontal direction. Near the edge the resulting stress can partially relax, as shown, by pushing away the material on the unimplanted side. (Displacements are exaggerated by a factor of ~100.)

the compressive stress on the implanted side to roughly half of its value far from the edge. For a negative magnetostriction constant, the effect of this local distortion is to cause an easy axis of magnetization to appear parallel to the pattern edge on the implanted side. On the other side of the implant boundary, of less immediate interest, the induced easy axis is normal to the edge.

Backerra et al. (1981), on the basis of observations of a different sort, suggested at about the same time as Hidaka and Matsutera that the stress-relief mechanism is important for the formation of charged walls at implant boundaries.

We will discuss the effects of the stress-relief-induced anisotropy near the implant edge partly following the treatment of Hubert (1984), who pointed out that not only does stress relief contribute essentially to the formation of charged walls at implant edges, but it also accounts for much of the threefold symmetric effects around an unimplanted disk that had traditionally been ascribed to the magnetocrystalline anisotropy.

Although a glance at Fig. 43 reveals that the stress and strain near the edge of an unimplanted disk vary in the direction perpendicular to the film in complex ways, with shear as well as compression being involved, our understanding of this process is insufficiently quantitative at present to justify a full three-dimensional treatment. Instead, we will consider only an effective two-dimensional stress in the plane of the film. The planar stress may be described by its principal values σ_1 along the principal axis parallel to the implant edge and $\sigma_2(<\sigma_1)$ and by the angle ω of the principal axis. Far from the edge, $\sigma_2 = \sigma_1$. The magnetoelastic energy is then (Hubert, 1984) given by

$$\begin{aligned}e_{ms} =& -\tfrac{3}{4}(\sigma_1 + \sigma_2)\lambda_{111}\cos^2\theta \\& -\tfrac{1}{4}(\sigma_1 - \sigma_2)(2\lambda_{111} + \lambda_{100})\cos^2\theta \cos 2(\phi - \omega) \\& + \sqrt{\tfrac{1}{2}}(\sigma_1 - \sigma_2)(\lambda_{100} - \lambda_{111})\sin\theta\cos\theta\cos(\phi + 2\omega).\end{aligned} \quad (9)$$

Here θ and ϕ are the polar and azimuthal angles of the magnetization, with $\theta = 90°$ being normal to the film, and λ_{111} and λ_{100} are the two independent magnetostriction constants of the cubic garnet. The first term in this equation is simply the uniaxial anisotropy that for $\lambda_{111} < 0$ and compressive stress ($\sigma < 0$) causes the magnetization to lie down in the implanted layer, and it is the only term that does not vanish far from the edge. The second term results in the easy axis along the implant edge that Hidaka and Matsutera (1981) showed to be necessary for the formation of charged walls. This circumferential easy axis causes the magnetization to align itself approximately parallel to the implant edge, an effect that was formerly credited only to the demagnetization energy. Hubert made a rough estimate of the latter and found it to be smaller than the stress relief effect. If we assume that the second term forces the magnetization in the implanted layer at the edge of the unimplanted disk to lie parallel to the edge ($\phi = \omega$), then the third term in the magnetostatic energy varies as $\cos 3\phi$, mimicking the threefold magnetocrystalline anisotropy energy but several times larger for typical values of the parameters involved.

Figure 42 showed the well-known "propeller domain" pattern around an unimplanted disk. The origin of this pattern has been discussed by Lin et al. (1978) and also by Callen (1979). The role of stress-relief anisotropy at the pattern edge had not been understood at that time; however, it happily turns out that the qualitative picture has not changed greatly, but has perhaps become more convincing as the result of the new understanding. One must merely realize that the magnetostrictive anisotropy that results in the easy axis parallel to the pattern edge is more effective than demagnetization effects in holding the magnetization along the edge, and that very near the edge the magnetostrictive anisotropy ($\lambda_{100} \neq \lambda_{111}$) overpowers the magnetocrystalline anisotropy in conventional materials.

Figure 44 (Callen, 1979) shows qualitatively how the propeller domains are formed near an unimplanted disk. At each point on the boundary of the disk, magnetostrictive and demagnetization effects favor the magnetocrystalline easy direction that is most nearly parallel to the edge. A low applied in-plane field to the left, however, suppresses the magnetization in the easy

Patterned Ion-Implanted Layers 269

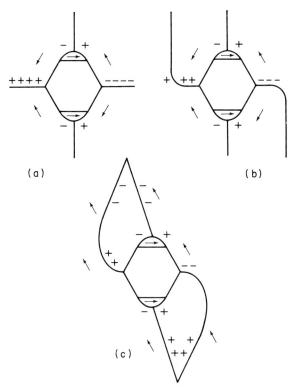

Fig. 44 Formation of propeller domains as shown in Fig. 42 under low in-plane fields. (a) low applied field to the left, (b) reduction of charged wall energy by bending, (c) closure of domains to favor one easy direction far from the disk. (From Callen, 1979.)

direction to the right. This gives us the picture of Fig. 44a: two charged walls of opposite sign project from the disk to the right and left and at the top and bottom there are uncharged walls. The long walls of maximum charge here are energetically costly, and in any case a reasonable boundary condition is that, far from the disk, the magnetization is everywhere in one or the other of the two easy directions favored by the applied field (we can assume that the applied field is a little closer to one of these directions than to the other). The charged walls will bend around to join the uncharged ones at top and bottom, and so we arrive finally at the classical configuration shown in Fig. 44c, which corresponds to the photograph in Fig. 42. The walls projecting to the right and left are now no longer uniformly charged; in fact the sign of the charge can change along the length of the wall as in this example. The fully charged parts of the walls are found only near the

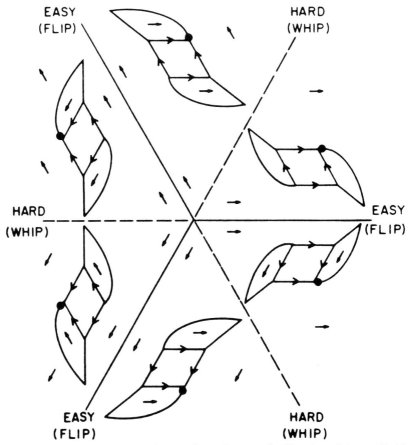

Fig. 45 Dependence of propeller configurations on the direction of the applied in-plane field. (From Callen, 1979.)

edge of the disk; away from the disks it is energetically good for them to turn rapidly to a direction of lower charge.

The propeller domains in Fig. 42 correspond to an in-plane field directed toward the left or a little upward of left. For other directions of applied field there will be similar patterns of different orientations, as shown in Fig. 45 (Callen, 1979). If we apply a rotating drive field, we will see the domain pattern go through these six configurations by means of the "whip and flip" transitions of Lin *et al.* (1978). Let us start with the configuration of Fig. 44, at the upper left in Fig. 45, and rotate the drive field counterclockwise. As the drive field crosses the indicated hard axis, the points at which the charged walls are anchored to the disk do not move far (a condition that is enforced primarily by magnetostrictive anisotropy at the pattern edge, as Hubert has shown), but they now curve in the opposite direction, and so "whip" from one direction to the other. A different sort of transition takes place when the next symmetry axis, an easy magnetization direction, is crossed. Now the points of attachment of the charged walls "flip" from one set of anchoring points to another. As the drive field sweeps around the circle, the charged walls whip and flip their way around also. We have given no details here about wall configurations during whips and flips, but photographs of transition configurations have been published (Dove and Schwarzl, 1979). In the low drive situation assumed here, the transitions obviously must involve walls sweeping over the whole sample, since the magnetization far from the disk switches direction when hard axes are crossed.

If we add a small "test bubble" (depicted by a black dot in the various parts of Fig. 45) to this picture, we would expect it to move around the disk in fits and starts as the point of attachment of the attractive charged wall flips and whips its way around the disk—if it can keep up! Real bubbles indeed behave this way for the lowest drive fields at which they can be made to propagate around a disk, spending most of their time near the anchoring points of attractive charged walls at the three hard directions. As the drive fields are increased, bubbles as well as the charged walls they are attached to are observed to move more smoothly around the disks than do charged walls in the absence of bubbles. Since drive fields are an order of magnitude lower than $4\pi M$ of the bubble layer, it is obvious that the bubble stray fields must present a major perturbation of the highly simplified picture of bubble propagation around a disk that we have described here.

Bubble propagation on contiguous disk tracks is of course a more involved process than the mere revolution around a disk described above.

7 DAMAGE AND STRAIN PROFILES IN THE IMPLANTED LAYER

The direct effect of ion implantation of magnetic bubble garnet is to damage the crystal structure, and the magnetic effects that are exploited for bubble devices are the results of this damage. We will here first consider the effects of implantation on the crystal and discuss some methods that have been developed to measure the consequences of the resulting damage.

A direct physical result of ion implantation is the swelling of the damaged region, due to the introduction of defects. Since the damaged layer is usually constrained by a thick substrate from expanding in its plane, the film remains stressed in the plane, and the strain resulting from ion implantation appears only in the normal direction.

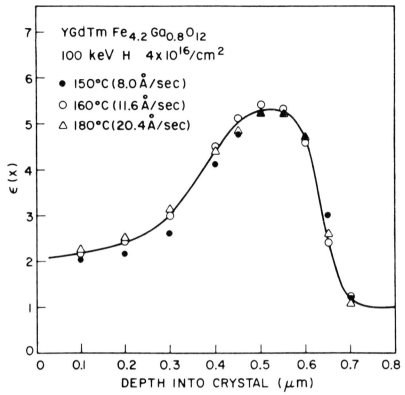

Fig. 46 Etching rate enhancement profiles for a 100-keV hydrogen-implanted bubble film showing the damage peak at 0.5 μm depth. The enhancement rates are temperature-independent. (From Johnson et al., 1973.)

A simple although destructive method of measuring the damage profile (or something closely related to it) of implanted garnet films resulted from the observation of Johnson et al. (1973) that the rate at which implanted garnet was etched in hot phosphoric acid is greater by a large factor than that of unimplanted garnet. They introduced the "etching rate enhancement factor" $\varepsilon(x)$, where ε is the ratio of the etching rates in implanted and unimplanted material and x is depth into the crystal, and showed that while the etching rate itself is strongly temperature-dependent, $\varepsilon(x)$ does not depend on temperature. For the 100-keV H implant used, they found that ε was approximately linear with dose in the range $1-4 \times 10^{16}/\text{cm}^2$. The etching rate enhancement profile in Fig. 46 shows that the damage due to the 100-keV H implant peaks strongly at a depth of $\sim 0.5\,\mu\text{m}$.

Perhaps the most direct way of measuring the strain in the damaged layer is by X-ray diffraction. The damaged layer, with a different lattice constant

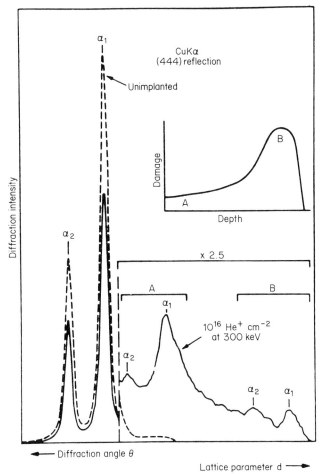

Fig. 47 X-ray diffraction curves of a bubble garnet before (dotted curve) and after (solid curve) implantation of 1×10^{16} He/cm^2. Peaks A corresponds to a region of low damage (and strain), while peaks B reflect the region of peak damage, as shown in the inset. (From North and Wolfe, 1973.)

(in one direction) from the underlying material, will exhibit satellite diffraction peaks shifted from the main peak due to the substrate by an amount proportional to the strain. The amplitudes of the satellite peaks carry information not only about the depth of the implanted layer but also about the extent of the damage. If the material is damaged to the point of amorphicity, the satellite peaks will of course vanish.

In practice, the damage due to implantation varies with depth into the film, and so does the resulting strain. Figure 47 (North and Wolfe, 1973)

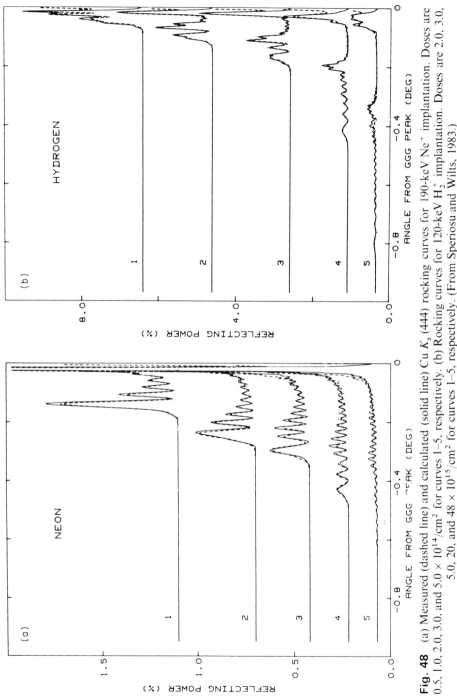

Fig. 48 (a) Measured (dashed line) and calculated (solid line) Cu K_α (444) rocking curves for 190-keV Ne$^+$ implantation. Doses are 0.5, 1.0, 2.0, 3.0, and $5.0 \times 10^{14}/\text{cm}^2$ for curves 1–5, respectively. (b) Rocking curves for 120-keV H$_2^+$ implantation. Doses are 2.0, 3.0, 5.0, 20, and $48 \times 10^{15}/\text{cm}^2$ for curves 1–5, respectively. (From Speriosu and Wilts, 1983.)

shows an example of the effect of the shape of the damage profile on low-resolution X-ray diffraction curves. While it is possible to obtain qualitative information about strain profiles from such measurements, greater resolution is needed to extract profiles in any detail.

With sufficient resolution, X-ray diffraction patterns will reflect the damage and strain profiles, but in a complicated way due to interference between reflections from atomic layers at various depths. Speriosu *et al.* (1979) were able to reconstruct strain and damage profiles from high-resolution double crystal diffraction measurements made on samples with successively more of the implanted material etched off. The profiles were built "from the bottom up" by first fitting the diffraction pattern of the thinnest layer and then adding new layers each time without changing the part of the distribution already determined. Speriosu (1981) has described the kinematic diffraction theory that is the basis of this method.

Direct inversion of the diffraction curves to find the damage and strain profiles is confounded by the fact that the phase of the diffracted radiation is not known. Nevertheless, if reasonable models for the distributions are assumed, it is possible to make fits to the diffraction pattern of the complete layer without resorting to etching it away in steps (de Roode and Smits, 1981; Speriosu and Wilts, 1983). While uniqueness of the results of this procedure cannot be proved, it is believed that only one strain distribution (except for a mirror reflection) will give a good fit to a measured diffraction pattern.

Figures 48 and 49 (Speriosu and Wilts, 1983) shows the quality of the fits that can be obtained by this method. Figure 48 shows the measured diffraction patterns obtained for several doses of monoenergetic Ne^+ and H_2^+ implantation in a bubble garnet material, as well as the corresponding calculated curves. The assumed strain profiles that yielded these excellent fits are shown in Fig. 49. The strain profiles for both H_2^+ and $Ne+$ are strongly peaked at definite depths in the material. The hydrogen implant (consisting in the material effectively of 60-keV protons rather than 120-keV H_2^+ ions) penetrates far deeper than the heavier $Ne+$ ions for a given energy, since the protons lose energy in electronic collisions less effectively. In general the strain profiles observed for ion implantion in magnetic bubble garnets agree well with predictions made from the LSS theory of nuclear ranges in matter. An interesting exception in the case of hydrogen will be discussed later.

The damage information yielded by the X-ray rocking curves is less precise than that obtained for strain. Speriosu and Wilts found that for the range of doses represented in Fig. 48 the area under the rocking curves decreases by about an order of magnitude with increasing dose. This is attributed to the decreasing structure factor $F(|F| = 0$ corresponds to amorphousness).

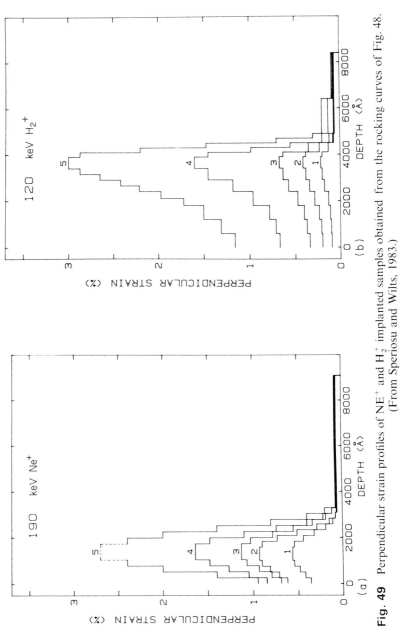

Fig. 49 Perpendicular strain profiles of Ne^+ and H_2^+ implanted samples obtained from the rocking curves of Fig. 48. (From Speriosu and Wilts, 1983.)

For strains $\lesssim 0.5\%$, the structure factor is not noticeably changed and no direct information is obtained on the damage profile. At a maximum strain varying from $\sim 2.5\%$ for neon to 3.9% for hydrogen the damage is so great and the structure factor so small that no more measurable diffraction occurs. Speriosu and Wilts find that the damage peak corresponds in depth to the strain peak, and that at the peak the standard deviation of the (assumed Gaussian) distribution of atomic displacements is proportional to strain. It is reasonable to assume that the same relation holds throughout, consistent with the assumption that damage is the source of the strain.

Another type of measurement that can provide profiles in implanted crystals is backscattering spectrometry of Mev ions. Paine *et al.* (1981) have compared this method to kinematic X-ray diffraction for use with garnets. The two methods agree on the strain at the surface, but diffraction provides a better strain versus depth profile. Damage profiles from the two methods agree to a factor of two, and neither as conventionally applied is sensitive to the distribution of light atoms. In view of the importance of the strain profile to bubble devices, the X-ray method is to be preferred for damage and strain measurements.

8 MAGNETIC PROPERTIES OF THE IMPLANTED LAYER

Although the garnet crystal structure is cubic, films intended for magnetic bubble devices are of course grown with a uniaxial anisotropy that favors the perpendicular orientation for the magnetization vector in the film, which allows them to support bubbles. Ion implantation changes the sign of the uniaxial anisotropy, producing a layer near the surface that has its magnetization vector in the plane of the film.

The primary mechanism by which ion implantation changes the uniaxial anisotropy is inverse magnetostriction. (Ordinary magnetostriction is the phenomenon by which the dimensions of a magnetic material change when it is subjected to a magnetic field.) In the inverse effect, a change of magnetic anisotropy results when the material is strained. The effect on the anisotropy energy K_u due to a strain $\Delta a/a$ perpendicular to the film plane is given by $\Delta K_u = 1.5 E \lambda_{111} v a/a(1 + v)$, where E is Young's modulus, v is Poisson's ratio, and λ_{111} is the relevant magnetostriction coefficient for the usual (111) film. For compression in the film plane to cause the magnetization vector to lie in the plane of the film, λ_{111} must be negative and sufficiently large to reverse the growth-induced anisotropy for a strain of $\sim 1\%$.

The trend in ion-implanted magnetic bubble devices is naturally toward smaller bubbles and higher storage density. Submicrometer bubble films will have higher $4\pi M$ and corresponding higher as-grown anisotropies. Will ion implantation be able to induce a sufficient anisotropy change to reverse the growth induced anisotropy of these materials? One approach to the implantation of small bubble materials that has been used at IBM (Lin et al., 1977) is to grow double-layer epitaxial films. The composition of the lower layer is tailored with large K_u for use as the storage layer, and the upper layer is grown with lower anisotropy to make it more readily implantable. Obviously the extra complexity of having to grow two epitaxial films per substrate is a considerable disadvantage with this technique, and it is desirable to stay with a single layer if possible.

The damage–strain distributions resulting from a single monoenergetic implant are strongly peaked at a specific depth. Uniform distributions would be expected to be more suitable for drive layers in ion-implanted bubble services, and Jouve et al. (1976) introduced the custom of making implants at more than one energy in order to flatten out the damage distribution.

Characterization and understanding of the effects of ion implantation on the magnetic properties of the implanted layer is a large and difficult task that is still in an incomplete state. Rather than just the uniaxial anisotropy, we must expect all of the magnetic parameters to be changed as a result of implantation damage (and in some cases perhaps by the presence of the resulting foreign ions), and in general the effects will again vary with depth into the film. Several techniques have been used for probing the magnetic structure of the implanted layer, including ferromagnetic resonance FMR, ac susceptibility measurements, conversion electron Mossbauer spectroscopy, vibrating sample magnetometry, and Kerr rotation.

It is of considerable interest to compare the magnetic effects of ion implantation with the primary mechanical effects of damage and strain that we have discussed earlier. Ferromagnetic resonance has proven to be a particularly useful tool for investigating the profiles of magnetic properties in the implanted layer and we will concentrate on what has been found out using FMR.

If the complete FMR spectrum is analyzed, a considerable amount of information concerning the profiles of magnetic properties can be extracted. The most complete development of this approach has been given by Wilts and Prasad (1981), who solved numerically the FMR resonance equations in films with a nonuniform surface layer. Their approach was to build up the magnetic profiles by fitting spectra of progressively thicker milled samples (in rough analogy to a method that has been used to derive strain profiles). Wilts and Prasad showed that for a satisfactory separation of the magnetic parameters that affect the resonance frequencies and amplitudes to be made, both perpendicular and parallel resonance measurements must be made. If the sample under consideration has an appreciable crystal anisotropy H_1, resonance measurements as a function of angle will also be needed in order to separate out its effects.

According to Wilts and Prasad, this complex but powerful method yields the profile of the uniaxial anisotropy $H'_k = H_k - 4\pi M$ to a few percent. Changes in the saturation magnetization M and the exchange constant A are not found so precisely, but if it is assumed that the profiles of these parameters are similar to that of H'_k, they may be found to $\sim 10\%$ in the implanted region.

A good comparison between strain and damage profiles on the one hand and magnetic properties on the other is provided by the work of Speriosu and Wilts (1983), who discuss three implant species: hydrogen, helium, and neon. We have already shown some of their strain profiles for H and Ne implants, obtained from X-ray rocking curves, in Fig. 49. The magnetic profiles obtained in this work come from the interpretation of FMR spectra. As in the interpretation of X-ray rocking curves, the analysis of these spectra begins with the assumption of trial distributions chosen with the aid of experience, and proceeds by trial and error. The process is more difficult than for X-rays due to the larger number of parameters involved.

Figure 50 shows representative perpendicular FMR spectra for the samples of Fig. 48. The major parameters that affect mode locations and amplitudes are the field for uniform resonance,

$$H_{un} = \omega/\gamma - H_k + 4\pi M + (\tfrac{2}{3})H_1, \tag{10}$$

and the ratio of magnetic stiffness to saturation magnetization A/M. When several modes are seen, as in the spectra of Fig. 50, there appears to be a unique relation between the FMR spectrum and the profile of H_{un}, provided that the H_{un} profile is unimodal. When this is not the case it is necessary to resort to removing the material in thin steps of 100–200 Å and reconstructing the profiles as described by Wilts and Prasad.

For "low" and "medium" doses of helium and neon, a unique relationship was found between the change ΔH_{un} due to implantation and the corresponding change in the strain $\Delta\varepsilon$. At low doses the relationship was linear, while at medium doses saturation of ΔH_{un} set in. For hydrogen implants no unique relationship was found, but rather the presence of the introduced hydrogen itself contributed to ΔH_{un}.

For high doses, ΔH_{un} becomes oversaturated and may even decrease again in the region of highest damage. An example is shown in Fig. 51, which compares ΔH_{un} to $\Delta\varepsilon$ for a high-dose helium implant of $1.2 \times 10^{16}/\text{cm}^2$. While the strain profile is still unimodal, the distribution of ΔH_{un} shows a dip suggesting a sort of magnetic overdamage in this region; $4\pi M$ and A/M both have drastic dips at about the same depth, as Fig. 51 shows. The value

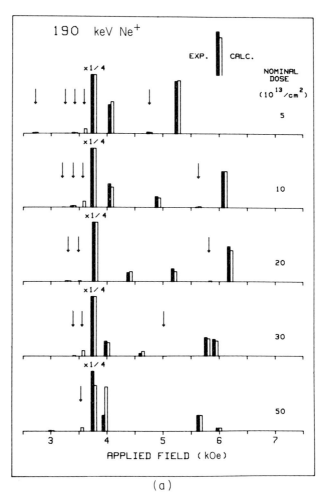

(a)

Fig. 50 Measured (solid rectangles) and calculated (open rectangles) perpendicular FMR spectra for the Ne$^+$ and three lower dose H$_2^+$ samples of Fig. 48. The actual measured fields are at the boundary between open and closed rectangles, whose heights are proportional to measured amplitudes. Actual mode linewidth increases with dose from 70 to ~150 Oe. The arrows indicate small amplitude experimental modes and theoretical modes of nearly zero amplitude with no experimental counterparts. (From Speriosu and Wilts, 1983.)

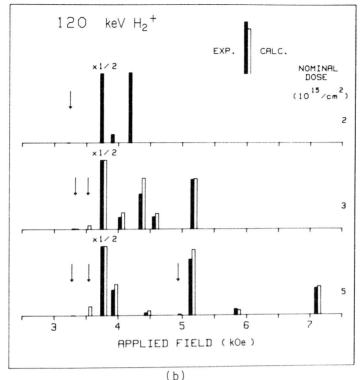

(b)

Fig. 50—contd.

of $4\pi M$ is reduced by 80% at the location of maximum strain, and A suffers a dramatic 98% reduction.

Figure 52 summarizes data from ΔH_{un} profiles for five doses of neon implantation. The value of ΔH_{un} is plotted against the corresponding value of delta strain, with different doses indicated by different symbols. For each dose there is a corresponding maximum strain that is indicated at the top. At low strains, points for all doses fall on the same curve with an initial slope of 4.1 kOe/%. (The same slope obtains in the case of helium, and in fact the rest of the corresponding diagram is qualitatively also very similar.) Speriosu and Wilts show from their estimates of the changes in other magnetic parameters that in the linear region 98% of ΔH_{un} is due to the change in uniaxial anisotropy H_k. If magnetostriction is assumed to be responsible for this change in H_k, then using bulk values of Young's modulus ($E = 2 \times 10^{12}$ dyne/cm^2), Poisson's ratio ($v = 0.29$), and magnetization ($4\pi M = 510$ G), the value $\lambda_{111} = -1.3 \times 10^{-6}$ is obtained for the magnetostriction coefficient, in reasonable agreement with the value -3.4×10^{-6} that

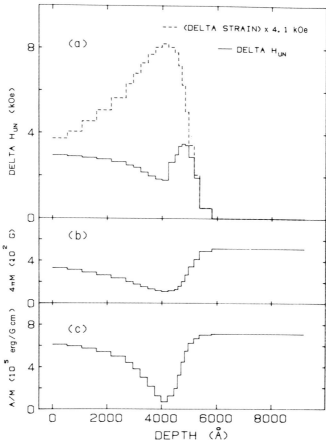

Fig. 51 Strain and magnetic profiles for a sample implanted with 1.2×10^{16} 140-keV He^+/cm^2, showing magnetic overdamage resulting from a heavy dose. The magnetic profiles were derived from FMR spectra made as a function of etch depth. (From Speriosu and Wilts, 1983.)

Speriosu and Wilts obtain from the material composition and published tables. This agreement and the fact that the same slope is found for low doses of He and Ne doses strongly support the notion that under these conditions magnetostriction is the mechanism by which ion implantation changes the uniaxial anisotropy.

For large strains, ΔH_{un} saturates and then decreases again to roughly zero at a strain where the material becomes paramagnetic. It is interesting although unexplained that in this region of heavy damage the relation between strain and ΔH_{un} is not unique. Rather, the material in the "inner region" deeper than the level of maximum damage is found to follow one

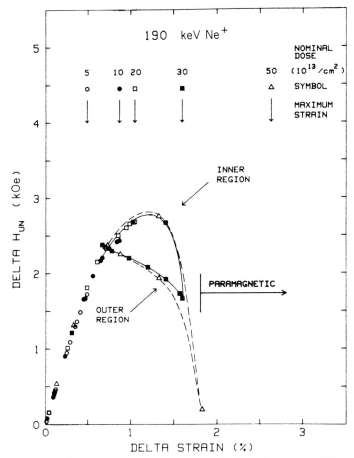

Fig. 52 Change in H_{un} versus change in strain for five Ne$^+$ doses. Different doses are represented by different symbols. The maximum strain obtained with each dose is indicated. (From Speriosu and Wilts, 1983.)

curve while the material above this point follows another. Results for helium implantation are very similar, except that the maximum ΔH_{un} is larger at about 3.5 kOe and occurs at a larger strain of ~1.5%, and the paramagnetic region also begins at a higher strain of 2.3%.

The behavior of the extremum values of the saturation magnetization $4\pi M$, exchange constant to magnetization ratio A/M, the cubic anisotropy field H_1, and the damping coefficient α are shown versus maximum strain for Ne implants in Fig. 53. Due to insufficient sensitivity of the FMR spectra on these parameters, their profiles cannot be determined, but good fits are obtained if the induced changes are assumed proportional to $\delta\varepsilon$ except that

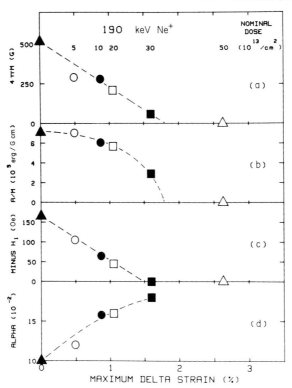

Fig. 53 Extremum values of magnetic properties vs change in strain for five Ne⁺ doses. Different doses are represented by different symbols. (a) Saturation magnetization. (b) Ratio A/M of exchange stiffness to saturation magnetization. (c) Cubic anisotropy field H_1. (d) Damping coefficient α. (From Speriosu and Wilts, 1983.)

$4\pi M$ and A/M are zero for strains greater than 1·8%. Again the diagram for helium implantation is very similar.

Hydrogen implantation is different. Figure 54, from Hirko and Ju (1980), shows the change in the FMR-measured anisotropy change [more properly, the change in the separation between the principal surface and body modes as Speriosu and Wilts (1983) pointed out] versus damage level for H⁺, He⁺, and B⁺ implantation, both before and after annealing at 350°C. In the preannealing data, the field change due to H⁺ implantation rises far above the 3000-Oe saturation level of the heavier ions. After annealing, the effect of hydrogen, although still larger than the others, is considerably diminished, while the effects of He⁺ and B⁺ are scarcely affected. Two important features of hydrogen implantation distinguish it from other species. First, since the nuclear stopping power of hydrogen is considerably less than that of other

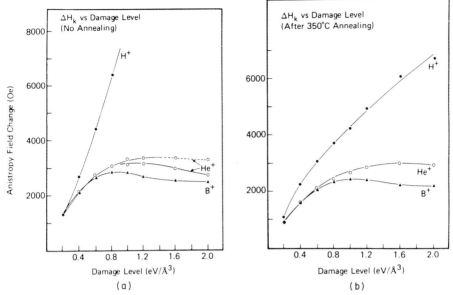

Fig. 54 Anisotropy field change versus implantation-induced damage level for H^+, He^+, and B^+ ions, (a) before and (b) after annealing at 350 C. A damage level of 1 eV/Å^3 corresponds very roughly to a change in strain of 1%. (From Hirko and Ju, 1980.)

ions, far larger doses of hydrogen are required to produce equivalent amounts of damage. This implies that if any chemical effects due to the presence of the dopant itself, as opposed to the damage it does to the host lattice, are present, these may be more noticeable in the case of hydrogen than for other species that render the garnet paramagnetic at far lower doses than required for hydrogen. The second distinguishing property of hydrogen is that it is very mobile in implanted garnet at higher temperatures, and desorbs from the implanted film rapidly at temperatures higher than 200°C (Sugita et al., 1981). Naturally, as the hydrogen leaves the crystal it will take with it any contribution to changes in magnetic properties due to its presence.

For an illustration of the anomalous (i.e., not due to normal magnetostriction) contribution of hydrogen to the change in uniaxial anisotropy, we turn again to the work of Speriosu and Wilts (1983). Figure 55 shows the ΔH_{un} profile obtained with a 120-keV H_2^+ dose of 5×10^{15} cm^{-2} compared to the contribution expected from magnetostriction, found by multiplying the implantation-induced strain by 4.1 kOe/%. The difference between these profiles, which is the anomalous part, is shown in Fig. 55c to agree remarkably well with the density of hydrogen atoms in the crystal calculated from LSS theory for 60-keV protons in garnet. The changes

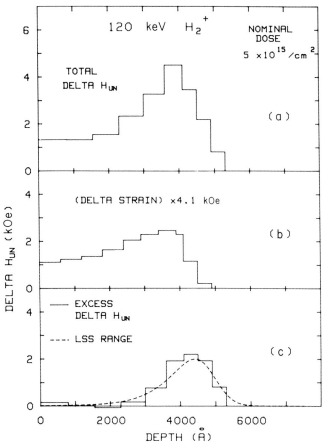

Fig. 55 (a) Depth profile of ΔH_{un} for implantation with $t \times 10^{15} H_2^+/cm^2$ at 120 keV, broken down into contributions due to (b) strain and (c) a chemical effect due to the presence of hydrogen. The calculated LSS range of 60-keV protons in garnet is shown by the dashed curve in (c). (From Speriosu and Wilts, 1983.)

induced in magnetic parameters other than H_{un} by hydrogen implantation are also unusual. If the same maximum strain of 0.6% in Fig. 55 is produced by Ne^+ or He^+, $4\pi M$ is reduced by less than 30% compared to the 60% reduction due to hydrogen. Hydrogen implantation was found by Speriosu and Wilts to increase the ratio A/M, while for neon and helium the exchange constant decreases faster than the magnetization so that A/M decreases.

Further evidence that the excess contribution to the change in H_{un} due to hydrogen implantation results from the presence of hydrogen is shown in Fig. 56 from Speriosu and Wilts, which compares total ΔH_{un} profiles with the expected magnetostrictive contributions for two hydrogen doses as

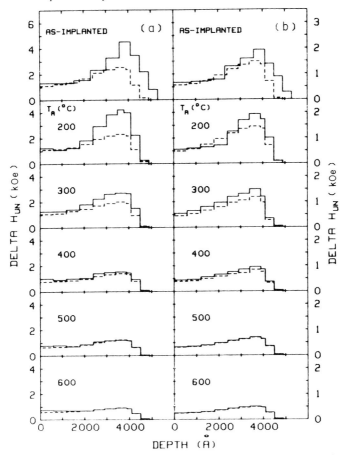

Fig. 56 Total ΔH_{un} profiles (solid) compared to $\Delta\varepsilon \times 4.1$ kOe profiles (dashed) for (a) $5 \times 10^{15}/\text{cm}^2$ and (b) $2 \times 10^{15}/\text{cm}^2$ 120-keV H_2^+ implantation. Annealing temperature increases toward the bottom. Note the different vertical scales for the two doses. (From Speriosu and Wilts, 1983.)

implanted and for various annealing temperatures. As the annealing temperature rises, two effects are evident. Due to healing of the damage, the strain-contributed ΔH_{un} decreases. Further, above 200°C the difference between the total ΔH_{un} and the magnetostrictive part falls off, with very little difference remaining at 400°C and above. The obvious interpretation is that the excess contribution due to the presence of hydrogen disappears as the hydrogen leaves the samples at higher temperatures.

From the work of Speriosu and Wilts and others referred to by them, it is evident that damage and strain in the implanted bubble film affect its magnetic parameters in essentially similar ways for implantation by different ions. For hydrogen in particular, additional chemical effects on the uniaxial anisotropy in excess of that produced by strain via magnetostriction as well as on other magnetic parameters are important. There is no reason to think that hydrogen as a dopant is unique in its chemical effects on magnetic properties, but similar effects may be difficult to detect with heavier ions. Speriosu and Wilts point out that for all other ions the implanted garnet becomes paramagnetic at doses far below those needed with hydrogen.

REFERENCES

Almasi, G. S., Giess, E. A., Hendel, R. J., Keefe, G. E., Lin, Y. S., and Slusarczuk, M. (1974). *AIP Conf. Proc.* No. 24, 630–632.
Almasi, G. S., Keefe, G. E., Lin, Y. S., and Sanders, I. L. (1980). *IEEE Trans. Magn.* **MAG-16**, 89–93.
Asada, S., and Urai, H. (1982). *Annu. Conf. Appl. Magn. Jpn., 6th,* Dig., Pap. 15pD-6.
Backerra, S. C. M., de Roode, W. H., and Enz, V. (1981). *Philips J. Res.* **36**, 112–123.
Best, J. S. (1981). *J. Appl. Phys.* **52**, 2367–2369.
Blank, S. L., Wolfe, R., Luther, L. C., LeCraw, R. C., Nelson, T. J., and Biolsi, W. A. (1979). *J. Appl. Phys.* **50**, 2155–2160.
Bobeck, A. H. (1983). *3IM3 Conf., Philadelphia, Pa.* Pap. FA-4.
Bobeck, A. H., Blank, S. L., and Levinstein, H. J. (1972). *Bell Syst. Tech. J.* **51**, 1431–1435.
Bonyhard, P. I. (1979). *J. Appl. Phys.* **50**, 2213–2215.
Bonyhard, P. I., and Nelson, T. J. (1982). *IEEE Trans. Magn.* **MAG-18**, 740–744.
Bonyhard, P. I., Hagedorn, F. B., Michaelis, P. C., Muehlner, D. J., Nelson, T. J., and Roman, B. J. (1980). *ICMB4, Tokyo* Pap. G-4.
Bonyhard, P. I., Hagedorn, F. B., Ekholm, D. T., Muehlner, D. J., Nelson, T. J., and Roman, B. J. (1982). *IEEE Trans. Magn.* **MAG-18**, 737–740.
Bonyhard, P. I., Ekholm, D. T., Hagedorn, F. B., Muehlner, D. J., and Roman, B. J. (1984). *IEEE Trans. Magn.* **MAG-20**, 129–134.
Bullock, D. (1977). U.S. Pat. 4,040,019.
Calhoun, B. A., and Sanders, I. L. (1981). *J. Appl. Phys.* **52**, 2386–2387.
Callen, H. (1979). *J. Appl. Phys.* **50**, 1457–1464.
Copeland, J. A., Josenhans, J. G., and Spiwak, R. R. (1973). *IEEE Trans. Magn.* **MAG-9**, 489–492.
de Leeuw, F. H. (1978). *IEEE Trans. Magn.* **MAG-14**, 596–598.
de Roode, W. H., and Smits, J. W. (1981). *J. Appl. Phys.* **52**, 3969–3973.
Dove, D. B., and Schwarzl, S. (1979). *J. Appl. Phys.* **50**, 5906–5913.
Ekholm, D. T., Bonyhard, P. I., Muehlner, D. J., and Nelson, T. J. (1982). *J. Appl. Phys.* **53**, 2525–2527.
Hidaka, Y., and Matsutera, H. (1981). *Appl. Phys. Lett.* **39**, 116–118.
Hirko, R., and Ju, K. (1980). *IEEE Trans. Magn.* **MAG-16**, 958–960.
Horng, C. T., and Schwenker, R. O. (1981). *J. Appl. Phys.* **52**, 2383–2385.
Hubert, A. (1979). *IEEE Trans. Magn.* **MAG-15**, 1251–1260.
Hubert, A. (1984). *IEEE Trans. Magn.* **MAG-20**, 1816–1821.

Ikeda, I., Takeuchi, T., Imura, R., Sato, T., Suzuki, R., and Sugita, Y. (1983). *Dig. Intermag Conf.* No. HA-3.
Johnson, W. A., North, J. C., and Wolfe, R. (1973). *J. Appl. Phys.* **44**, 4753–4757.
Jouve, H., and Puchalska, I. B. (1979). *IEEE Trans. Magn.* **MAG-15**, 1016–1020.
Jouve, H., Segalini, S., and Piaguet, J. (1976). *IEEE Trans. Magn.* **MAG-12**, 660–661.
Ju, K., Hu, H. L., Hirko, R. G., Moore, E. B., Saiki, D. Y., and Schwenker, R. O. (1981). *IBM J. Res. Dev.* **25**, 295–302.
Kleman, M., and Puchalska, I. B. (1980). *J. Magn. Magn. Mater.* **15**, 1473-1476.
Komenou, K., Ohashi, M., Miyashita, T., Matsuda, K., Satoh, Y., and Yamaguchi, K. (1981). *IEEE Trans. Magn.* **MAG-17**, 2908–2913.
Komenou, K., Matsuda, K., Miyashita, T., Ohashi, M., Betsui, K., Satoh, Y., and Yamagishi, K. (1982). *IEEE Trans. Magn.* **MAG-18**, 1352–1354.
Kryder, M. H. (1979). *IEEE Trans. Magn.* **MAG-15**, 1009–1016.
Kryder, M. H. (1981). *IEEE Trans. Magn.* **MAG-17**, 2385–2392.
Lin, Y. S., and Sanders, I. L. (1981). *IEEE Spectrum* **18**, 30–34.
Lin, Y. S., Almasi, G. S., and Keefe, G. E. (1977). *J. Appl. Phys.* **48**, 5201–5208.
Lin, Y. S., Dove, D. B., Schwarzl, S., and Shir, C. C. (1978). *IEEE Trans. Magn.* **MAG-14**, 494–499.
Lin, Y. S., Almasi, G. S., Dove, D. B., Keefe, G. E., and Shir, C. C. (1979a). *J. Appl. Phys.* **50**, 2258–2260.
Lin, Y. S., Almasi, G. S., Keefe, G. E., and Pugh, E. W. (1979b). *IEEE Trans. Magn.* **MAG-15**, 1642–1647.
MacNeal, B. E., and Gergis, I. S. (1981). *J. Appl. Phys.* **52**, 2380–2382.
Marsh, J. S. (1938). "The Alloys of Iron and Nickel", McGraw-Hill, New York.
Nelson, T. J., Chen, Y. S., ands Geusic, J. E. (1973). *IEEE Trans. Magn.* **MAG-9**, 289–293.
Nelson, T. J., Wolfe, R., Blank, S. L., and Johnson, W. A. (1979). *J. Appl. Phys.* **50**, 2261–2263.
Nelson, T. J., Wolfe, R., Blank, S. L., Bonyhard, P. I., Johnson, W. A., Roman, B. J., and Vella-Coleiro, G. P. (1980). *Bell Syst. Tech. J.* **59**, 229–257.
Nelson, T. J., Bonyhard, P. I., Geusic, J. E., Hagedorn, F. B., Johnson, W. A., and Wagner, W. D. P. (1981). *IEEE Trans. Magn.* **MAG-17**, 1134–1141.
Nelson, T. J., Ballintine, J. E., Reith, L. A., Roman, B. J., Slusky, S. E. G., and Wolfe, R. (1982). *IEEE Trans. Magn.* **MAG-18**, 1358–1360.
Nelson, T. J., Fratello, V. J., Reith, L. A., and Roman, B. J. (1983). *Dig. Intermag. Conf.* No HA-1.
North, J. C., and Wolfe, R. (1973). *In* "Ion-Implantation in Semiconductors and Other Materials" (B. L. Crowder ed.), pp. 502–522. Plenum, New York.
North, J. C., Wolfe, R., and Nelson, T. J. (1978). *J. Vac. Sci. Technol.* **15**, 1675–1684.
Paine, B. M., Speriosu, V. S., Wielunsk, I. L. S., Glass, H. L., and Nicolet, M. A. (1981). *Nucl. Instrum. Methods* **191**, 80–86.
Sanders, I. L., Kabelac, W. J., and Keefe, G. E. (1982). *IEEE Trans. Magn.* **MAG-18**, 745–749.
Satoh, T., Ohashi, M., Miyashita, T., Matsuda, K., Betsui, K., Komenou, K., and Yamagishi, K. (1982). *IEEE Trans. Magn.* **MAG-18**, 1355–1357.
Satoh, Y., Komenou, K., Miyashita, T., Ohashi, M., Betsui, K., Matsuda, K., and Yamagishi, K. (1983). *Dig. Intermag Conf.* No. GA-1.
Shir, C. C. (1981). *J. Appl. Phys.* **52**, 2388–2390.

Shir, C. C., and Lin, Y. S. (1979). *J. Appl. Phys.* **50**, 2270–2272.
Speriosu, V. S. (1981). *J. Appl. Phys.* **52**, 6094–6103.
Speriosu, V. S., and Wilts, C. H. (1983). *J. Appl. Phys.* **54**, 3325–3343.
Speriosu, V. S., Glass, H. L., and Kobayashi, T. (1979). *Appl. Phys. Lett.* **34**, 539–542.
Sugita, Y., Imura, R., Takeuchi, T., Ikeda, T., Suzuki, R., Ohta, N., and Umezaki, H. (1981). *Magn. Magn. Mater. Conf., Atlanta, Ga.* Pap. BA-1.
Suzuki, R., Takeuchi, T., Kodama, N., Takeshita, M., Ikeda, T., and Sugita, Y. (1982). *3IM3 Conf., Montreal* Pap. FB-01.
Tabor, W. J., Bobeck, A. H., Vella-Coleiro, G. P., and Rosencwaig, A. (1972). *Bell Syst. Tech. J.* **91**, 1427–1431.
Urai, H., and Asada, S. (1982). *Annu. Conf. Magn. Jpn., 6th,* Dig., Pap. 15pD-4.
Urai, H., and Yoshimi, K. (1982). *Annu. Conf. Magn. Jpn., 6th,* Dig., Pap. 15pD-6.
Urai, H., Morimoto, A., and Mizuno, K. (1982). *Annu. Conf. Magn. Jpn., 6th,* Dig., Pap. 15pD-2.
Vella-Coleiro, G. P., Wolfe, R., Blank, S. L., Caruso, R., Nelson, T. J., and Rana, V. V. S. (1981). *J. Appl. Phys.* **52**, 2355–2357.
Wilts, C. H., and Prasad, S. (1981). *IEEE Trans. Magn.* **MAG-17**, 2405–2414.
Wolfe, R., and Nelson, T. J. (1978). *Intermag '78, Florence* Pap. II-1.
Wolfe, R., and Nelson, T. J. (1979). *IEEE Trans. Magn.* **MAG-15**, 1323–1325.
Wolfe, R., and North, J. C. (1972). *Bell Syst. Tech. J.* **51**, 1436–1440.
Wolfe, R., and North, J. C. (1974). *Appl. Phys. Lett.* **25**, 122–124.
Wolfe, R., North, J. C., Barns, R. L., Robinson, M., and Levinstein, H. J. (1971). *Appl. Phys. Lett.* **19**, 298–300.
Wolfe, R., North, J. C., Johnson, W. A., Spiwak, R. R., Varnerin, L. J., and Fischer, R. F. (1973). *AIP Conf. Proc.* No. 10, 339–343.
Wolfe, R., Muehlner, D. J., and Nelson, T. J. (1981). *J. Appl. Phys.* **52**, 2377–2379.

Index

Access time, 143, 178, 182
Al-Cu, 83, 143, 185
Alignment, 76
Alloys, 41
　crystalline alloys, 49
Amorphous, 41, 71
　amorphous alloys annealing, 55
　amorphous alloy device performance, 78
　amorphous alloys ion implantation, 57
　amorphous alloys magnetic anisotropy, 59
　amorphous alloys magnetizations, 50
　amorphous film fabrication, 34
　amorphous films domain wall dynamics, 66
　amorphous materials, 31
Anisotropy, 34, 41, 52, 59, 62, 106, 109, 120, 126, 158, 174, 225
　anisotropy energy, 265
　cubic anisotropy, 92
　growth induced anisotropy, 107
　in-plane anisotropy, 113, 126, 174
　magnetocrystalline anisotropy, 265, 267
　magnetocrystalline anisotropy energy, 268
　shape anisotropy, 147, 164
　stress-induced anisotropy, 126
　uniaxial anisotropy, 53, 92, 217, 265, 268, 277
Argon, 52, 59
Asteroid, 247
Atomic arrangement, 41
Atomic order, 46
Au, 75

$BaFe_{12}O_{19}$, 93
Baffles, 9
Bi_2O_3, 127
Bismuth, 92, 128
　bismuth-containing films, 94, 127
$(Bi, Y, Ca)_3(Fe, Si)_5O_{12}$, 129
Bit, 155, 182
Bloch line, 113, 116
Bloch wall, 95, 113, 115
Bohr magneton, 50
Boltzmann constant, 50
Bubbles, 59
　A-type bubbles, 116
　B-type bubbles, 116
　bubble diameter, 52, 116, 165
　bubble translation experiments, 121
　bubbles static properties, 113
　influence of in-plane fields on the static properties of bubbles, 116
　temperature dependence of bubble parameters, 129
Bubbles dynamic properties, 120, 128
Boundary, 216

Cascade, 58
Cation, 20
Ceramic, 18
Chevron, 71, 80, 81, 144
　asymmetric chevron, 166
Chip, 208
　chip organization, 183, 224
　1-Mbit chip, 184
Circular loops, 17
Cluster, 59
Co, 33, 41, 56
Coercivity, 24, 53, 93
Coil, 131, 184

291

Collapse field, 53
Column, 47
Compensation point, 62, 63
Compensation temperature, 49, 52
Conductor, 131, 142, 149, 174, 185, 247, 149
Coordination, 3
　co-ordination number, 43
　octahedral coordination, 3
　tetrahedral coordination, 3
Cracks, 4, 56
Cross-section, 58
Crucible, 9, 10, 11
Crystal, 4, 204, 271, 285
　crystal stoichiometry, 18
　crystal defects, 4
Crystallization, 34, 55, 78
Cu, 56
Curie temperature, 49, 52, 103, 105, 112
Current, 76, 80, 83, 121, 132, 143, 149, 157, 252
Cusp, 235, 247, 249, 257, 259
Czochralski growth, 3, 4, 8, 18

Damage, 58, 272, 279, 287
Damping, 126, 128, 283
Data, 159, 189, 226
Data rate, 181
Demagnetization field, 95, 112, 266
Demagnetizing energy, 59, 61, 116
Detection, 251
Detector, 81, 138, 141, 161, 227
　Nelson detector, 163
　serpentine detector, 163
Device, 34, 53, 75, 138, 224, 238
　amorphous film bubble devices, 75
　contiguous-disk device, 133, 195, 216, 219
　current access devices, 130
　high density device, 139
　ion-implanted devices, 194
Dielectric breakdown, 76, 77, 81, 84
Dielectric film, 81
Diffusion, 56
Dipole, 65
Dipolar induced anisotropy, 65
Direction cosine, 65
Dislocations, 2, 10, 12, 98
　linear dislocations, 12

Domain motion, 10
　domain wall damping, 125
　domain wall mobility, 53
　domain wall motion, 94
　planar domain walls, 263
　propeller domain, 268
　straight domain wall experiments, 120
Drive field, 52, 78, 140, 152, 168, 187
　minimum drive field, 140
Dynamic properties, 66, 93

Eddy currents, 74
Electromigration, 83, 84, 143, 153, 175
Electron diffraction, 41, 47
Electron density, 46
Electron scattering, 59
　chemical effects, 288
　function elements, 143
Epitaxy, 2
Error, 154, 157, 161, 190, 201
$Eu_3Ga_5O_{12}$, 22
$(Eu, Lu)_3(Fe, Al)_5O_{12}$, 97
Evaporation, 45, 56, 60
Exchange energy, 53
Exchange constant A, 53, 92, 109, 112, 221, 283
Exchange field, 33
Exchange interactions, 33

Facets, 5
　facet formation, 5
　facet region, 7
　facet strain, 5
Faraday rotation, 94
Fe, 33
Feature,
　minimum feature size, 142
Ferromagnetic resonance, 65, 94, 125, 128, 278
Field,
　field access, 130
　mean field, 55, 62
　molecular field, 103
Flip, 63, 271
Fluid, 15
Frequency, 78, 95, 131, 142, 152, 182, 252

Index

Ga, 99, 104, 108, 109, 127
Ga_2O_3, 11
Gap, 142, 166, 195, 216, 234
Garnet, 56, 71, 76, 93, 97, 128
   gadolinium gallium garnet, 2, 3, 4, 5, 6, 30, 100, 127
   gallium garnets, 22
   rare-earth aluminium, 2
   rare-earth iron garnets, 1, 2, 106
   rare-earth gallium garnets, 2, 3
Gas, gas pressure, 36, 37
Gas content, 36, 37
Gd, 103, 109
   Gd–Co, 34, 47, 67, 68
   Gd–Co–Au, 67, 68
   Gd–Co–Cu, 67, 68
   Gd–Co–Fe, 60
   Gd–Co–Mo, 67, 68
   Gd–Fe, 47
   Gd_2O_3, 12
   $Gd_3Ga_5O_{12}$, 3, 22
   $Gd_3Sc_2Al_3O_{12}$, 26
   $(Gd, Y)_3(Fe, Mn, Ga)_5O_{12}$, 97, 99, 102, 105, 107, 108, 113, 114, 123
Generator, 157, 245
GGG *see* Gadolinium gallium garnet
Growth interface, 7, 13
Growth striations, 6, 26
Gyromagnetic ratio, 51, 105
Gyroscopic property, 94
   gradient, 124
   groove, 175

Hall coefficients, 32, 63
Hall effect, 63
Hard-sphere model, 43
Helical defects, 15
Helical dislocation, 16
Helix, 15
Hillocks, 98
   Helium implantation, 283
   Hydrogen implantation, 284
   Neon implantation, 281
Hysteresis loop, 71

Implantation dose, 57
Impurity, 5

Inclusions, 2, 10
   Cubic inclusions, 11
   Iridium inclusions, 11
Insulator, 34, 82
Interface, 13
   solid–liquid interface, 6
   seed-crystal interface, 13
Ions, 21, 48, 57, 59, 97, 103, 126
   ion implantation, 55, 56, 75, 205, 216
   ion substitution, 21, 24
Iridium, 11

Kerr effect, 50

La, 99
Length parameter, 52, 93
   single-mask-level, 75
   multi-mask level, 77
Lattice parameter, 2, 3, 6, 7, 18, 105
Lattice mismatch, 4
Length parameter, 112, 221
Lift-off, 78
Lithography, 194, 219
$Ln_3Sc_2Ga_3O_{12}$, 3, 27
$Ln_3Sc_2Al_3O_{12}$, 3, 27
$Ln_3Ga_5O_{12}$, 27
Loop, 78, 240, 259
   closed loop, 17
   loop organization, 178
   major loop, 231, 258, 262
Lu, 127

Magnetic anisotropy, 4, 57, 59, 60, 216
Magnetization, 34, 53, 92, 103, 108, 112, 120, 155, 257, 268, 277, 283
Magneto-optic, 71
Magneto-resistive effect, 78, 161
Magnetostriction, 60, 106, 107, 109, 163, 175, 225
Manganese, 99, 100, 107
   $MnCo_3$, 100
MnO_2, 100
Mn_2O_3, 100
Mask, 75
Melt, 3, 98, 127
Microdiffraction, 45

Mismatch, 107–109
Mo, 52
Mobility, 67, 74, 78, 125, 126, 128, 131

$Nd_3Ga_5O_{12}$, 22, 28, 128
Néel wall, 113, 115
Neutron diffraction, 41
Neutron scattering, 33
Ni, 33
NiFe, 75, 77, 83
Nitrogen, 60
Noise, 141, 164, 188, 190
Nuclear resonance, 33
Nucleation, 116

Orthoferrite, 93
Orthorhombic anisotropy, 92, 95, 97, 99, 106, 115
Overshoot, 122
Oxidation, 56
Oxygen, 33, 35, 60, 63, 100
  oxygen segregation, 6
  oxygen vacancy, 5, 6

Page, 191
Pair, 65
Pattern,
  pattern period, 167
  "wide-gap" patterns, 167
Pb, 105
PbO, 127
Permalloy, 75, 130, 133, 138, 165, 175, 185, 193, 200, 210, 216, 225, 255
Perpendicular magnetic anisotropy, 35, 47, 60
Photolithography, 142, 194
Photoresist, 219
Pickax, 149
Planar anisotropy, 35
Plasma, 35
PLOS, 185, 209
Point defects, 10
Pole, 61, 141, 147, 149, 157, 166, 195
Precipitates, 15
Production, 194
Propagation, 256
Propagation margins, 202

Proton, 275
Pt, 105

Quality factor, 53, 92

Radial distribution function, 41
Radices, 59
Rare-earth materials, 32
$Re_3Sc_2Ga_5O_{12}$, 25
Registration, 142, 176
Replicator, 80, 148
Replication, 249
Resin, 175
Resistivity, 32, 34
Rod, 61

Scandium, 24
Scattering, 47
  small angle scattering, 47, 63
Sensitivity, 81
Shape effect, 62, 64
Shuttle, 225, 258
SiO, 184
SiO_2, 71, 82, 175
Signal, 81, 190
Site,
  octahedral site, 20, 21, 103, 106
  dodecahedral site, 20, 21, 103, 106
  tetrahedral site, 20, 21, 106
Skew angle, 123
$Sm_3Ga_5O_{12}$, 13, 22, 28
$Sn_3Ga_5O_{12}$, 13, 22, 28
Solid solution, 23
Solubility, 18
Spacing, 145, 170, 175, 197, 203, 206
Spin, 65
  spin wave, 33
Sputtering, 35, 56, 59
  sputtering targets, 37
Step, 15
Stochiometry, 5
Strain, 2, 5, 6, 10, 116, 216, 271, 279, 287
Stress, 4, 15, 107, 109, 131, 174, 216, 266
Stretcher, 144
  chevron stretcher, 145
  stretcher margin, 146

Index 295

Striations, 7, 9
Strip domain, 115
Strip domain under influence of in-plane fields, 115
Structural anisotropy, 44, 45, 61
Structure, 49, 59
Subnetwork, 50
Suboxyde, 11
Supercooling, 98
Susceptibility, 63

TaMo, 184
Target,
  arc-melted target, 35
  hot-pressed target, 35
TbFe, 47
Teflon, 131
Temperature,
  annealing temperature, 287
Temperature range, 142
Track,
  contiguous disk track, 259
  crystal orientation effect on wide-gap track margins, 204
  half-disk tracks, 166
  roof-top track, 262
  track orientation, 256
  "wide-gap" track, 200, 202
  wide-gap track propagation margins, 199, 210
  wide-gap track spacing thickness dependence, 206
Transfer,
  field effect transfer, 234
  gradient effect transfer, 235
  reverse rotation transfer, 242
  transfer gates, 234
  transfer gates/swap gates, 153

Transfer—*contd.*
  transfer switch, 77, 80
  trapping transfer, 239
Trap, 239

Vacancy, 56
Vacuum, 55
Velocity, 67, 80, 94, 120, 122, 124, 128
Void, 46, 56, 61
Volatility, 231
Voltage, 143, 153, 182, 188

Wafer, 142, 158, 171, 175, 184
Wall,
  charged walls, 263, 269
  charged-wall formation, 266
  wall damping σ_w 93
  wall energy, 93
  wall pinning, 93
Waveform, 161
"Whip", 271
"Wide-gap", 196

X-ray diffraction, 41, 275
X-ray technique, 12
X-ray scattering, 59

Y, 99, 106, 127
YAG, 8
Yield, 143
YIG, 6, 105, 106, 109, 129, 223
$YFeO_3$, 95
$(YGdTm)_3(FeGa)_5O_{12}$, 217
$(YPb)_3(FeMn)_5O_{12}$, 102
YSmLuCaGeIG, 152, 165, 184, 209